Old Heart of Nevada

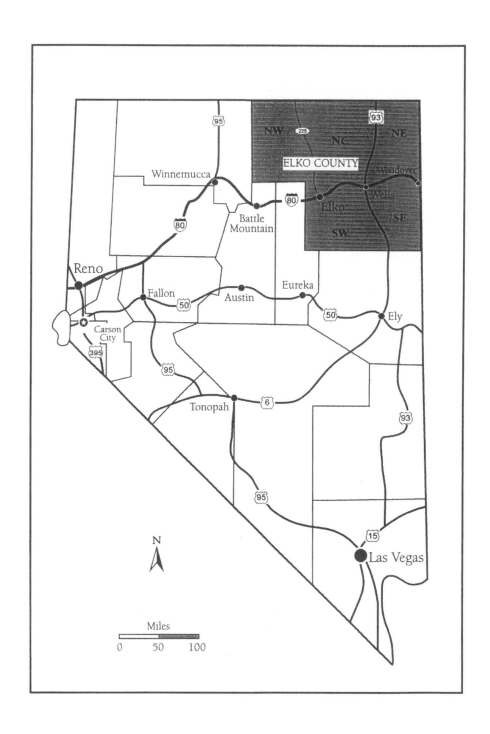

ELKO COUNTY

NW
224
NC
93
NE
Wendover
Winnemucca
80
Wells
Elko
Battle
Mountain
SW
SE

95

Reno
Fallon
Eureka
Austin
50
50
Ely

Carson
City

395

95

6
Tonopah
93

95

N

15
Las Vegas

Miles
0   50   100

Old Heart of Nevada

*Ghost Towns and Mining Camps
of Elko County*

SHAWN HALL

University of Nevada Press   Reno / Las Vegas

University of Nevada Press, Reno, Nevada 89557 USA

Copyright © 1998 by University of Nevada Press

All rights reserved

Manufactured in the United States of America

Library of Congress Cataloging-in-Publication Data

Hall, Shawn, 1960–

Old heart of Nevada : ghost towns and mining camps of

Elko County / Shawn Hall.

p.   cm.

Includes bibliographical references (p.   ) and index.

ISBN 0-87417-295-0 (pbk. : alk. paper)

1. Ghost Towns—Nevada—Elko County—Guidebooks.

2. Mining camps—Nevada—Elko County—Guidebooks.

3. Elko County (Nev.)—History, Local.   I. Title.

F847.E4.H35 1997

96-49497

917.93'160434—dc21   CIP

The paper used in this book meets the requirements of the

American National Standard for Information Sciences—

Permanence of Paper for Printed Library Materials,

ANSI Z39.48-1992 (R2002). Binding materials were

selected for strength and durability.

15   14   13   12   11   10   09   08   07

5   4   3

To my "Pop"

A L B E R T   W I L L I A M   H A L L

It was very special that we were able to share ghost town adventures in Elko County during the preparation of this book.

    His support and faith in me have provided the inspiration to overcome the obstacles and challenges encountered as I progress down the road of life.

Contents

Preface

When I started researching this book in 1986 I had not envisioned the depth and breadth of the undertaking. Frequently, while I was working on the project, I thought I had finally finished the research and was ready to begin writing, only to find additional sources of information. The writing phase entailed transforming well over 1,000 pages of notes into an organized, readable format. Finishing the task took eight years. Of all the books I have written, this one has been the most challenging, because of its huge scope. It initially covered almost 300 historic sites, ghost towns, stage stations, railroad towns, ranches, mining camps, and forts, but the size of the volume was unacceptable to the publisher. I have thus pared it down so that it includes only a history of Elko County's ghost towns and mining camps. While people still live in some of the towns this book covers, these sites are included because they satisfy the definition of a ghost town: a shadowy semblance of a former self. The many stage stations, railroad stops, ranches, valleys, and other historical sites beyond the scope of this book will be discussed in a later volume.

Traveling the almost 15,000 miles it took to reach and every site mentioned in the book was an altogether enjoyable experience. At times, Elko County's beauty overwhelmed my camper's heart. It wasn't until I moved to Elko that I was able to truly enjoy all the wonders the county has to offer.

I have tried to present as accurate a history as possible of each site in Elko County. Despite the intense scrutiny that the manuscript has undergone by myself and local history buffs, errors may be present. I ask any reader who comes across questionable information to contact the author so that corrections can be made in future editions.

Many historical sites in Elko County have been degraded over the years. I have witnessed much of this inexcusable vandalism firsthand. Buildings have been destroyed by people trying to find out if there is anything inside their walls. Foundations and sites have been ruined by overzealous bottle hunters. Anyone who witnesses such vandalism should let the proper authorities know about it. I hope that my children will have the opportunity to enjoy the historic sites of this county as I have, but unless local residents, visitors, and federal agencies show some respect, there will be nothing left to see.

A good number of the sites discussed in this book are on private property. Please honor all private property signs and contact the owner before exploring them. A simple request for permission to visit is often all the property owner wants. I hope others enjoy visiting Elko County's historic spots as much as I have. While I no longer live in Elko, I treasure the memories acquired during my four years there. The remembrances I choose to keep strong in my mind are the positive ones.

Acknowledgments

First and foremost, I have to thank my lovely daughter Heather Ashley Hall. The happiest moments of my life are when I am with her, and it was such a pleasure to have her with me while visiting many of the places in Elko County. Without Heather, my life would be empty.

I would also like to offer my special thanks to my parents, Albert and Lorraine Hall. They continue to be supportive of my endeavors. It is very difficult to live almost 3,000 miles away from them.

Special thanks go to Howard and Terry Hickson. Through Howard's guidance, I was able to learn about the intricacies of museum management. I am deeply indebted to both him and Terry for their friendship. I also thank my friends in the history field for their assistance and support: Bob and Dorothy Nylen, Wally Cuchine, Bill Metscher, Gloria Harjes, Phillip Earl, and Doug and Cindy Sutherland. My best friend, Bruce Franchini, continues to be an important part of my life. While we see a lot less of each other these days, playing mountain Frisbee golf and camping with him are among my most memorable moments.

Many other people have been instrumental in making this book a reality, both through their friendship and through their assistance with historic information. They are: BeBe Adams, Matile Griswold Allen, Hazel Anderson, Delmo Andreozzi, the Barrick II city league basketball team, Dan and Dana Bennett, Betty Bear, Robin Boies, Louise Walther Botsford, Bob Burns, Lois and Charlie Chapin, Lee and Mary Chapman, Arthur Clawson, the late William E. Clawson, Jr., PJ Connolly, Desert Sunrise Rotary, Alan Duewel, the Elko Chamber of Commerce and its board of directors, Rick Fleishmann, Fred Frampton, Helen Fullenwider, Cliff Gardner, Norman and Nelda Glaser, Meg Glaser, Steve Glaser, Lee Hoffman, Art and Cora Holling, Wilma Holloway, Lori Kocinski, Gregg Lawrence, Irene Davis Linder, Lorry Lipparelli, Peter Marble, Carole and Frank Martin, Joe McDaniel, the late Dr. Leslie Moren, Joe Nardone, Dan and Karen O'Connor, Dennis Parks, Edna Patterson, Jim Polkinghorne, Tony and Ellen Primeaux, Elizabeth Pruitt, Jerry Reynolds, Diane Beitia Rice, Pauline Riordan, Paul Sawyer, Jane and Ray Schuckman, Gertrude Sharp, Bob Songer, Anne Steninger, Ed Strickland, Christina Ulm, Paul Walther, Jack and Francis Wanamaker, Barbara Wellington, Max Wignall, Helen Wilson, and Bob Wright. I will always treasure the true friendships that I forged during my years in Elko, and I thank those people who made me feel welcome there.

Northwestern Elko County

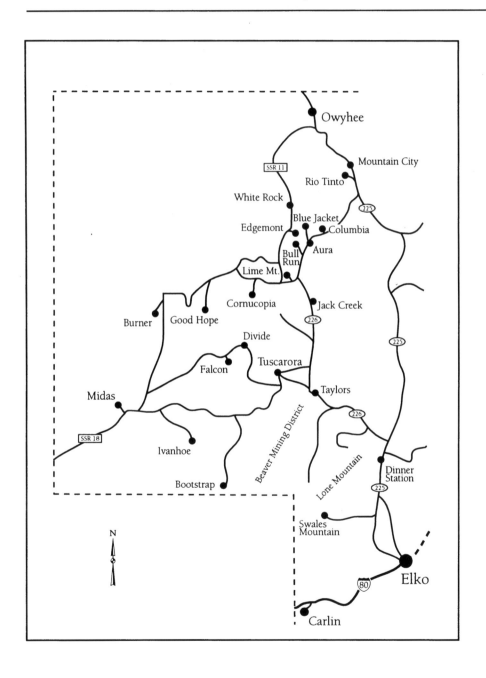

Owyhee

Mountain City

SSR 11

Rio Tinto

White Rock

225

Blue Jacket

Edgemont   Columbia

Bull   Aura
Run

Lime Mt.

Cornucopia   Jack Creek

Burner   Good Hope   226

Divide   225

Falcon   Tuscarora

Midas   Taylors

SSR 18   226

Ivanhoe

Bootstrap

Dinner
Station

Swales   225
Mountain

N   Elko

80

Carlin

Aura

DIRECTIONS: *Located 1½ miles south of Columbia.*

The revival of mines in Blue Jacket Canyon and at Bull Run led to Aura's formation in late 1905 when a townsite was platted in March 1906. The name Aura comes from *aurum,* Latin for gold. The town was located at the junction of the Mountain City Road and the road heading to mines in Blue Jacket Canyon. While Aura boomed, nearby Columbia, which had been empty since the turn of the century, did not revive. On July 5, 1906, a post office opened in Aura. Barney Horn, who was active in local freighting and mining interests, replaced postmaster A.L. Womack within a couple of months. Horn also ran the mail from Tuscarora to Mountain City via Deep Creek and Aura.

When the Aura Mining District was organized in 1907 it encompassed Blue Jacket, Bull Run, and Columbia—areas discussed elsewhere in this book. There was no mining at Aura, but the town was the center of what was once known as the Centennial Mining District. Aura served as a supply depot for local mining companies and miners and was a stop on the western branch of the Northern Stage Company, which ran from Tuscarora to Mountain City. The fare was four dollars. Population at Aura and the surrounding mines grew from 20 in early 1906 to 150 by 1907. As long as mining in the area was productive, Aura remained alive.

By 1907 businesses, including a saloon, two stores, and two boarding-houses, were active. The post office was housed in a sturdy stone building.

There are accounts of a newspaper called the *Concentrator* in town, but they are incorrect. The only paper by that name was published at Betty O'Neal (in Lander County). A school opened in 1907, and by 1915 it had grown large enough to need a principal. The school was active until the Aura School District was abolished in July 1927.

In February 1911 the Aura post office was robbed. While away on business, postmaster Barney Horn had left a boy named Higuera in charge, and Higuera stole a number of postal money orders. He was caught when he tried to cash them, because he had filled them out incorrectly. In 1914 Gus Dryden replaced Horn as postmaster. Dryden served until the office closed on September 30, 1921.

By 1917 the bulk of the mining activity in the Aura District was over, and most people had left. In 1920 nothing but some leasing activity was going on. The town of Aura emptied. All businesses were gone by time the post office closed, and no one ever came back to live at Aura. Some of the remaining wood-frame buildings were moved elsewhere, and only the stone walls of the Aura Saloon are left. While remains are scant at Aura, there are many spots to visit in the area. The drive to the town offers some of the best scenery in Elko County. High rocky mountains and heavily forested valleys surround the site.

Beaver Mining District
(Lake Mountain)

DIRECTIONS: *Located 18 miles south of Tuscarora on the old Carlin–Tuscarora stage road.*

Some small gold placer deposits were discovered on Dip Creek in 1872, and the Beaver Mining District was organized. During the placer operation, large amounts of float turquoise were recovered. The turquoise lode developed into the Stampede mine and began producing. The district is reputed to have generated revenue in the amount of $100,000 during the ensuing fifteen years, but newspaper accounts of production, intended to promote the mines, are misleading. When placer deposits ran out by the 1890s the Stampede mine produced turquoise intermittently until the 1960s, bringing the total value of materials mined to $500,000. However, barite became the biggest commodity mined in the Beaver Mining District when the Lakes mine was discovered in 1959, and soon the Beacon and Coyote mines began production as well. In 1980 a small concentrating jig plant was built, but deposits were running out and by 1983, after 700,000 tons of barite were produced, all the mines had shut down. Today, open pits and trenches mark the site.

Blue Jacket
(Blythe City) (Milford)

DIRECTIONS: *From Aura, take poor dirt road heading directly north for 3 miles to Blue Jacket.*

A prospecting party led by Jesse Cope discovered the Blue Jacket mine in 1869. Afterwards, Cope figured prominently in copper discoveries in the Mountain City area. Blue Jacket was the first mine discovered in the Centennial Mining District, which the Aura Mining District later absorbed. Shortly after Cope's discovery, other prospectors filed claims along Blue Jacket Canyon and nearby Columbia Creek. A town that provided virtually all the supplies for mines and miners developed at Columbia. By July 1871 five mining companies were working mines in Blue Jacket Canyon. They were the Blue Jacket, the Johnson, the Ontario, the Nevada, and the Fifteenth Amendment.

The Blue Jacket mine was by far the most productive mine in the area, and a small camp, named Blythe City after the owner, Thomas Blythe, formed nearby. In October 1875 a ten-stamp mill was completed a mile below the mine alongside Blue Jacket Creek, which was water-powered and equipped with Bruckner furnaces. About fifteen people lived at the mill and another twenty-five resided at the mine. Mine owners built a tramway, used off and on for forty years, from the mine to the mill, because the road to the mine was too steep and dangerous for ore wagons. By 1877 Blythe City had grown enough that a justice of the peace and a constable were named.

Blythe died in 1879, and his property became part of the Blythe estate. Without his leadership, however, the mine never regained its prominence. In 1880 English interests bought a number of mines in the Centennial District from the Blythe estate. These mines included the Blue Jacket, the Tuscarora, the Pioneer, the Hoosier State, the Ontario, the Revenue, and the Emigrant. Joe McClay built a five-stamp mill, the Cumberland, below the old Blue Jacket mill. The Cumberland and Argent water ditches were dug to bring additional water to the mill, but the venture proved a dismal failure.

In April 1885 James White bought the Blue Jacket mine and mill, still an active producer, for $85,000. In July of that year James Dawley overhauled the mill, which had been idle for six years. He restarted it in July, with William Milford taking over the managerial duties. By August, sixty men were working at the mine and mill, and White started a tunnel intended to come out on the White Rock side of the Centennial Range. In September Blythe City was renamed Milford after mill superintendent William Milford, but for reasons still unclear, the company curtailed all operations in October and abandoned the property, despite the fact that it was shipping more than $1,000 worth of gold and silver ore a week. Sixty men were left without jobs, and the property reverted back to the Blythe estate.

This strange turn of events ended many productive years for the Blue Jacket mine and mill, though the mine revived in 1889, and various prospectors worked it until the early teens. The refurbished tramway brought ore down to trucks. New stock was issued, and the Blue Jacket Mining Company formed. The Blythe estate, which was still mired in a bitter dispute in the San Fran-

Old mill located halfway up Blue Jacket Canyon. (Photo by Shawn Hall)

cisco courts, leased the mine and mill to the Blue Jacket Mining Company. Florence Blythe, a long-lost daughter from England, showed up claiming the estate, and rumors spread that the town would be renamed Florence City. Testimony in the case finally concluded in April 1890. In August the courts declared Florence Blythe the estate's sole heir, but she had no interest in mining, and there was little chance of a revival. The mill, while impressive, failed because the district's ore was difficult to treat, and only a security guard and a sheep dog lived at the mill's camp.

But the Blythe Estate saga was far from over. In December 1892 the Supreme Court of California affirmed that Florence was the sole heir of Thomas Blythe's millions. She was also responsible for legal fees in the amount of $834,000. In the meantime, the mine and mill had been attached by Thomas Trezine, who was owed $16,300, and the property was sold at a sheriff's auction. However, little was done with the property and the mill, and by the teens they had fallen into disrepair and were unusable.

More productive mining came to lower Blue Jacket Creek in 1903, when the Walker brothers discovered the Aura King mine. The Aura King Gold Mining Company, which owned the Aura King, the Del Monte, the Tecumseh, the Humboldt, and the Gold Bug mines, was incorporated in November, with R. P. Hunter as president. During the next three years, two five-stamp mills started operations in the canyon. In 1906 the Blewett brothers discovered the Jackpot mine and purchased the Walker mill, built in July 1904. By 1908 two shifts were working the Jackpot mine. The Aura King Company erected buildings including two boardinghouses, a mess hall, shops, and the

Aura King Hotel to serve its miners. The company mill ran twenty-four hours a day, crushing ore the mine produced. The Aura King Number One was the biggest ore producer.

In July 1906 the Denver-Nevada Placer Mining Company began working 2,200 acres of ground. H. L. Morris supervised the construction of a 5,000-foot flume, but the operation had limited success before folding in 1908. In August 1909 the Walker brothers built a small cyanide plant above the Aura King camp. Continued production of and interest in Blue Jacket properties led to the creation of the town of Aura, located a mile away. Mines' and miners' supplies were shipped through Aura. By this time Columbia had been dead for years and did not benefit from the Blue Jacket activity.

In 1908, Blue Jacket Canyon's best year, the mines produced almost 12,000 ounces of silver and 160 ounces of gold. A strike made in April 1912 in the Bonanza Jackpot, owned by the Blewett brothers, John Taylor, A. W. Sewell, and Barney Horn (postmaster at Aura), proved a valuable one. The Walker brothers' cyanide mill treated the ore, but while a number of mines continued to produce through 1917, amounts produced tailed off drastically after 1914. The Aura King Company folded in 1917, and all the mills closed. Since the 1920s only occasional leasing attempts have been made in the area.

Remains in Blue Jacket Canyon make it one of the more fascinating places to visit in Elko County. At the Aura King camp, one clapboard boardinghouse stands amid the collapsed remains of four other buildings. Mining equipment and parts litter the site. The road is completely washed out a couple of hundred yards above the ruins, which necessitates a strenuous hike up the picturesque canyon to reach Blue Jacket. Luckily, there are a number of wonderful campsites at Aura King to use as a base camp.

About a mile up the canyon are the ruins of a small five-stamp mill that was last used in the 1950s. Further up the creek, where the road turns to head up to the Blue Jacket mine, are the remains of the ten-stamp mill, which consist of stone foundations, crushing equipment, and scattered bricks from the stack. A couple of large wooden wheels from the tramway are also left, and evidence of the line exists all the way to the mine. Besides collapsed transoms, parts remain, including some with dates from the early 1870s.

A number of smaller mines and dumps exist halfway to the Blue Jacket mine. The mine, located near the top of Blue Jacket Peak, reveals large mine dumps and a couple of buildings. The mine building is a relic of the revival that took place after the turn of the century. The Blythe City townsite is now buried under the mine dump, and the only remaining structure from the 1880s is a collapsing outhouse. Just above the mine are remnants of the original Blue Jacket mine. Stone foundations mark the beginning of the tram and large concrete pads with huge bolts show where the line was anchored. Remains of metal piping and wooden flume sections are evidence of Blythe City's water supply, and the strongly flowing spring still runs. While only sheep and wildlife occupy the canyon and creek today, there are signs that extensive exploration has taken place during the past few years. Drilling roads crisscross the sides of Blue Jacket Canyon. Whether mining returns to this area remains to be seen, but the extensive mining remnants are mute testimony that gold and silver still exist here. For more information on the long mining history of the Centennial or Aura District, see the sections on Aura, Bull Run, and Columbia.

Bootstrap
(Boulder Creek)

DIRECTIONS: From Tuscarora, head south for 14.8 miles. Turn left and follow the road for 15 miles to Bootstrap.

Frank Maloney, who was employed by the Modoc Exploration Company, discovered the Bootstrap mine in 1940. Previous exploration in the area in 1914 had led to the establishment of the Boulder Creek Mining District when some antimony was mined, but the operation was abandoned after one shipment was made. The Modoc Company filed six claims, and soon Bootstrap developed into one of the first open pit mines in the United States. In 1949 the Getchell Mines Company leased the property and spent $14,000 in development, but the company gave up in 1950, and the Home-stake Mining Company leased the property in 1955. Marion Fisher, of Battle Mountain, formed a partnership with Harry Treweek (Gold Acres) and Robert Taylor (Battle Mountain) and took over the lease. The three men became the prime movers in the Bootstrap mine's development.

Ray Reed of Fallon began construction of a 100-ton mill on November 17. Ore was treated first by leaching, then by precipitation, and was then roasted to concentrate the gold. From 1957 to 1960 the mine produced almost 10,000 ounces of gold. Although nothing was produced during the next few years, a small crew stayed on at the mine. An employee burning trash in July 1964 caused a fire that ravaged 30,000 acres, most of which were owned by the T Lazy S Ranch. The next year, the ranch won a $136,000 judgment against the Bootstrap mine, and all the property was sold at public auction in February 1966.

The Newmont Mining Company, armed with its new microscopic gold extraction technique, acquired Bootstrap in 1967. After five years of development, production began in 1974. Ten years later, after it had produced about 120,000 ounces of gold, the deposit was mined out. However, the district is still active, with the Dee mine located about one mile northwest of the Boot-strap mine, and Newmont's Capstone deposit adjacent to the old Bootstrap mine. The mines along the Carlin Trend are among the richest in the world, and now produce well over three million ounces of gold per year. Barite is also an important mineral in the district. The Rossi mine, four miles north of Bootstrap, produced two million tons of barite between 1937 and 1989. Baroid Drilling Fluid Company ran the mine from 1947 to 1967, and Tom Norris, a mining contractor from Battle Mountain, ran it from 1967 to 1989. Norris built a jig mill, which operated until 1983. Baroid once again began working the property after Norris left, and the Mine Service and Supply Company refurbished the jig mill and restarted it in November 1993. Crushed ore is hauled to Dunphy for bagging and shipment.

Bull Run

DIRECTIONS: *From Aura, head south for 3 miles. Park here and hike west for 3 miles to Bull Run.*

The discovery of the Bull Run mine high on a mountainside on June 13, 1869, led to the formation of a small camp. Because of inaccessibility and low grade ore, however, the mine produced nothing, and little additional work was done there until near the turn of the century. New gold discoveries made in June 1896 lured people from Tuscarora to Bull Run. Among the first of the new arrivals were Steve Ferguson and Matt Graham, who staked a number of claims. Graham also had mining interests near Mardis. In January 1898 John Murphy, whom the *Tuscarora Times-Review* nicknamed the "Bull Run Bonanza King," reported that Bull Run was flourishing. In the same month, John Murphy, Frank Curieux, and Andrew Smith formed the Bull Run Gold and Silver Mining and Milling Company. Smith was the former owner of the Brewery Saloon in Tuscarora. The company also worked claims on Lime Mountain.

In August 1899 H.R. Cooke and his partners purchased a small mill that James Dove and W.S. Hillman, the Tuscarora undertaker, had built in the spring. The mill was dismantled and moved to the Cooke claim, which later developed into the Big Bob mine. In August Smith completed a small cyanide mill. Construction began on a ten-stamp mill adjacent to the cyanide plant in 1899. The mill, which cost $3,200, had to be moved from the old Rescue Mining Company property near Charleston. Lew Bernard tore down the mill and a large crew built a road to Bull Run. Fourteen teams moved the mill to Bull Run.

In October 1899 C.D. Roberts, who worked in the reconstruction of the mill, arrived at home and noticed the dirt roof of his log cabin was leaking. He went outside and threw some more dirt on the roof. To his horror, it collapsed on his wife and two children. All were hurt but none as badly as Roberts's young son, Vernile, who remained trapped under the dirt for fifteen minutes before rescuers could dig him out. Dr. Francis Drake came from

White Rock, but there was little he could do, and Vernile died from a skull fracture. He was buried in Tuscarora.

The mill, a unique self-loading and self-dumping style designed by H.C. Messimer, was completed and fired in November 1900. J.L. Powell was in charge of the stamp mill and Messimer ran the cyanide plant. Both had formerly been in charge of the Dexter mill and cyanide plant in Tuscarora. In May 1901 C.E. Woodward and W.J. Eastman sold their Hidden Treasure mine to the Montana Limited Mining Company of Marysville, Montana, for $6,000. In October the company hired William Strickler to add a concentrating unit to the mill. At the same time, some bunkhouses and a boardinghouse were built. In March Dan McKenna built a brick building and retort furnace for the assay office of the Murphy-Curieux Company. A small camp formed near the mill at an altitude of 8,000 feet on the only flat area at Bull Run. This not only meant harsh weather conditions but also extreme difficulty in hauling in mining and food supplies. The cost of hauling materials, particularly fuel for the mill, cut heavily into profits.

Frank Curieux's brother, Mortimer, of Tuscarora formed a partnership with Murphy. The Murphy-Curieux Mining Company worked a number of mines from 1900 to 1903, and the ore produced was processed at the mill, which was known as the Murphy-Curieux mill. The two companies were closely intertwined, and it was sometimes difficult to distinguish between them. Andrew Smith, part owner of the Bull Run Company, served as superintendent for the Murphy-Curieux Company. Mortimer Curieux was an assayer by trade and served both companies. The summer of 1901 began one of the best

Future director of the Bureau of Land Management, Marion Clawson, at his family's restaurant, Bull Run, 1906. (W.E. Clawson, Jr. collection, Northeastern Nevada Museum)

Old Heart of Nevada

Two miners' cabins still stand at Bull Run. (Photo by Shawn Hall)

periods in Bull Run history. Five thousand dollars worth of ore was shipped in July, $9,000 worth in August, $9,000 worth in September, and $8,000 worth in October. With the boom came a new necessity. Walt Hurley opened a saloon and a store, but in August 1902 the mining companies, in an attempt to control their miners, bought him out.

In December Mortimer Curieux, Andy Smith's brother-in-law, married Lillian Johnson in North Fork. The *Tuscarora Times-Review* made note that justice Henry Regan served as executioner. After Curieux married Johnson, he brought his new wife to live at Bull Run and the Curieuxs were one of only a few families with their own home at the camp. A wooden flume transported water from springs located half a mile away. After it failed one winter, it was replaced by a metal pipeline. A number of bunkhouses, boardinghouses, and small cabins accommodated the thirty residents. In September the mill ended a continuous fifteen-month run before it needed repairs. The value of bullion shipments totaled $100,000, and plans were made to bring in electric power to increase output.

In January of 1903 the Leonard Taylor Mining Company of Albany, New York, purchased the Bull Run Mining Company's and the Murphy-Curieux Mining Company's property, including the mill, for $150,000. For recreation, residents organized a baseball team and played against Edgemont, Tuscarora, and Elko. By 1903 the mines had almost a mile of underground workings. Reports put the value of production at as high as $120,000. In 1904 the company hired William Clawson, who had worked as mining superintendent in many places including Edgemont and Blue Jacket, to take charge of the mining and milling operations. He moved to Bull Run with his wife, Agnes, who, fortunately for historians, was an avid photographer and took pictures of places like Bull Run, Blue Jacket, and Aura. Her photographs are the only existing vintage views of the area. When the Clawsons moved to Bull Run, forty men were living at the camp, and Agnes was the only woman.

The Taylor Company closed down its operations in 1907, and the Clawsons left and moved to serve a new mining company at Blue Jacket. Most of the other residents also left, and by 1910 only a handful of determined miners still worked the mines. Occasionally enough ore was stockpiled to reactivate the mill for a short run. Consistent production never occurred, and only $20,000 worth of ore was mined from 1909 to 1920. The Bull Run District was silent until the mid-1930s, when the Centennial Gold Mining Company began working the Bull Run mine and built a twenty-five-ton mill in July of 1936. In August of 1939 a 100-ton Gibson mill began production, but operations were curtailed in 1940 after the mine produced $37,000 worth of ore. This was Bull Run's last bout of production. Some attempts at production were made in the 1940s and 1950s, but no ore was shipped.

Today, Bull Run is difficult to reach, but spectacular views and interesting remnants make seeing it well worth the effort for the approximately four-mile hike. The access road has been fenced off to help protect Bull Run Reservoir. Enjoy the first part of the hike, because the last mile is extremely steep. The first signs of Bull Run are remains of the wood flume and pipeline. It is easy to understand, while hiking to Bull Run, why it was so difficult to bring in supplies and haul out ore. At Bull Run, a couple of the bunkhouses still stand, although the boardinghouse/kitchen recently collapsed. Ruins of many other buildings are scattered on the small plateau. Mill remains are about a quarter of a mile past Bull Run. The ruins of the ten-stamp mill are the most prominent. While the stamps are gone, heavy timbers and the frame are still there. Until two or three years ago, the mill from the 1930s revival stood perched on the edge of the cliff, but avalanches later swept it into the canyon below. Avalanches also knocked the stamp mill frame off its foundations.

The ambitious adventurer will enjoy a couple of mines located near the crest of the mountain. For tamer explorers, the Bull Run mine is just above the mill ruins. Views from Bull Run are spectacular. From where the camp stood, the cliff drops off about 1,000 feet.

Old cabin at Burner.
(Photo by Shawn Hall)

Burner

(Burner Hills) (Elite Mining District)

DIRECTIONS: *From Midas, head east for 6 miles. Exit left and follow the road for 18 miles. Turn left and follow the road 1½ miles to Burner.*

Elijah and J.F. Burner made the first discoveries at Burner in December 1876. Assays ran as high as $400 per ton, and a load of ore was shipped to be treated at the Leopard mill at Cornucopia. The Burners called their main discovery the Viola Virginia mine. Other mines they established were the Julia, the Priscilla, the Victor, the Neva May, the King Imanuel, and the Grace Darling. The district was initially called Elite but was soon renamed Burner.

By April 1879 ten men were working a number of claims in the district, but little production occurred until Burner and his son, Victor, discovered the Mint mine. In January 1884 the Burners found a new vein that resulted in the development of the Garland King and the Neva mines. In 1884 Gil and Will Strickler, who also had interests in Bullion and Bull Run, found a small lode, but gave it up when the vein pinched out after limited production. S. L. Baldwin worked the Cleveland in 1885 but had little success. A. M. McAfee, J. F. Burner, and Will Strickler opened the Ella and Red Rock mines. The Burners worked their mines until 1893. Others gave the Burner District a try. During the summer of 1889 McAfee worked a number of claims.

In 1892 Strickler came back with Wellington Dove, a prominent Tuscarora mining man, and they hauled ore showing values in silver, gold, lead, and iron via Iron Point to the Carlin Sampling Works. Before the vein disappeared in late 1893, more than $15,000 worth of ore had been shipped to the Selby Smelter in Salt Lake City. Strickler tried again in 1897 in partnership with A.L. Snider, the Tuscarora postmaster, but little came of the venture. From 1877 to 1893, $30,000 worth of silver, gold, lead, and iron was produced.

Little if anything occurred in the Burner District until the 1930s when the Mint mine reopened. In 1936 a thirty-ton ball and flotation mill began producing ore. It was a small operation that remained active until the early 1940s. It produced ore valued at less than $20,000. After the 1940s the district saw no more production. Some exploration occurred in the early 1960s, but nothing developed. During the exploration, mine owners moved a trailer next to the Mint mine to house drillers. Recently, companies looking for disseminated gold deposits have conducted exploration in the area. It is possible that someday soon Burner could come alive again.

Nothing much was ever built at Burner, since there were never was more than fifteen people there. One bunkhouse from the 1930s, a trailer from the 1960s, and the foundations of the mill remain. The entrance of the Mint mine, which is flooded, is located next to the bunkhouse.

Columbia
(Van Duzer) (Marseilles) (Pennsylvania Hill)

DIRECTIONS: *From Aura, continue north for 1½ miles. Turn left and follow the road for ½ mile to Columbia.*

Columbia came into being in 1868 as a result of discoveries in Blue Jacket Canyon and on Maggie Creek. Further exploration led to additional discoveries at the boomtown of Columbia. Jesse Cope, one of the early discoverers, also found the rich Cope District, later called Mountain City. The Nevada mine was the first discovery at Columbia, and soon the Porter, the Potosi, the Montana, the Brigadier, the Monument, the Buster, the OK, the New Light, the Johnson, the Tuscarora, and the Ontario mines became active. By July 1870 these mines had produced $25,000 worth of ore, and twenty-five mines were operating. The *Elko Independent* declared that the district would soon take a position among Elko County's prominent mining camps.

In 1871 the Columbia Consolidated Mining Company organized and became the driving force behind Columbia's development. A twenty-stamp mill soon went up, but its efficiency level was less than 60 percent. Because all other facilities were far away, however, the mill was used extensively. Before it closed in 1883 the mill produced more than 40,000 tons of ore, for a value of $1.5 million. The Columbia Company owned the Infidel and the Big Four mines, two of the Columbia's best producers. Fishermen discovered the Big

Four when they spotted a silver ledge after a snow slide. In August 1871 the Found Treasure and Monument mines sold for $32,000. By the end of 1871 a camp of 100 had formed at Columbia, but the seventy-five-mile rough road separating Columbia and Carlin hindered the town's growth.

In October 1875 Elias DeYoung and James Duncan discovered the Mountain Laurel mine north of Columbia, and the Marseilles Mining District organized, with Duncan as recorder. In June 1877 two other mines, the Silver Reef and the Gallinipper, became active in the Marseilles District. Other mines opened during the next couple of years and included the Great Relief, the Eagle, the Infallible, and the Inexhaustible. A camp of fifteen formed at the mines.

In January 1880 the Columbia Company bought all properties in the district for $65,000 and the Marseilles Mining District became part of the Centennial Mining District. During the 1870s and 1880s as many as 200 Chinese worked the Van Duzer placers and other deposits in nearby streams. In 1895 the placers revived, and the Pennsylvania Hill Mining District organized in 1896 when miners removed $54,000 worth of placer gold. In addition, Dean and Dick Young were working the Gold Bug group, Ferdinand Testariere was driving a tunnel in nearby California Gulch, and Walter Stofiel was mining the Midnight group. While the placers had faded by 1900, Testariere's perseverance paid off when he struck a vein of ore worth $300 per ton in May 1902. It was only a small pocket, however, and he gave up working the mine in 1903. While the success of placer mining varied greatly, up to this day miners continue to work the deposits sporadically.

In 1875 Edward Stokes built a new mill in Columbia. It worked ore from the Revenue and Infidel mines, but it was unsuccessful and shut down in 1877. By 1878 Columbia's population had peaked at 125. The town consisted of two stores, two hotels, and eighteen homes. A dentist, Dr. W. W. Parsons, practiced there. The town became a stop on the Northern Stage Company's western branch, and the fare from Tuscarora to Columbia was four dollars. Clay Irland, a Columbia resident, drove the twice-a-week stage, carrying mail and passengers.

The Infidel mine and mill, under the supervision of Joel Meacham, employed most of the eighty miners in the district. Other producing mines at this time included the Maggie, the Comet, the Thomas Paine, the Bonanza, and the Bonanza Queen. A post office opened on February 24, 1879, with Thomas Smith as postmaster, and a school was built in 1880.

By 1881, however, many of the smaller mines had shut down. Columbia subsequently suffered a devastating blow when the Columbia Consolidated Mining Company folded in March 1883. In June E. M. Chapin leased the mill, which he restarted in July. It worked ore from the Leopard and Hussey mines in Cornucopia. In December W. C. Price, superintendent of the Navajo mine in Tuscarora, spent $3,000 and purchased all the property formerly owned by the old Columbia Company, including the mill, water rights, ditches, stone buildings, and the Columbia, the Infidel, the Maggie, the Hidden Hand, the Bonanza, the Bonanza Queen, and the Mountain Laurel mines. Brash newspaper reports stated that Price and his company took $300,000 out of the Maggie mine, but records of such glories do not exist. A school was built on Trail Creek in the 1880s, but in 1892, after the Maggie mine closed, it became part of the Cope School District.

In June 1884 Price, along with partners George Peltier, A. V. Lancaster, E. L. McMahan, and J. W. Powell, incorporated the Commercial Mill and Mining Company. Some of these partners also formed the Polaris Consolidated Mining Company to work the newly discovered Polaris mine, and in September Joseph McClay and James Clark located the Golden Eagle mine. Despite this activity, Columbia was fading fast, and while the town still had a justice of the peace and a constable, the Jarvis Store and the Can Can Restaurant (which later turned into a saloon) were the only businesses left. The school continued to operate until 1894, when there were no more students.

The folding of the Commercial Company in 1886 doomed Columbia. Just eleven voters participated in the November election of that year, and Columbia's mines produced only small amounts of gold and silver during the next fifteen years. By the time the post office closed on September 15, 1902, only a few people still lived in the town.

The district awoke in 1902 when new discoveries were made at Bull Run and Blue Jacket, but Columbia never revived. A new town, Aura, prospered a couple of miles south of Columbia. In its productive years from 1867 to 1902

Columbia mines produced 60,000 ounces of gold and four million ounces of silver. During this time, Columbia was the second most productive silver camp in Elko County, behind Tuscarora. In 1983 the Homestake Mining Company began exploration in Wood Gulch, a few miles north. Its mine began producing in 1988, but were mined out in late 1989 after yielding 49,000 ounces of gold and 200,000 ounces of silver. Mill foundations, stone ruins, and a couple of buildings remain near the Columbia Ranch, nestled in one of the most heavily forested areas of deciduous trees in Nevada. For more information on this area, see the sections on Aura, Blue Jacket, and Bull Run.

Cornucopia
(Milltown) (Mill City) (Kaufmanville)
(Baldinsville) (Brownsville)

DIRECTIONS: *From Jack Creek, head north on Nevada 226 for 5.8 miles. Exit left and follow the road for 9½ miles to Cornucopia.*

Cornucopia saw only a few years of activity, but while it existed it was a prosperous boomtown at the forefront of Elko County production figures. Martin Durfee, referred to in some writings as Matt, discovered the first ore there in 1872. He took his samples to Jack Coffman, a friend and assayer in Columbia. Durfee and Merrill Freeman made the first two claims, the Chloride and the Leopard, and the boom was on.

By January 1873 there were already seventy miners in what was known as Kaufmanville. One of the first businesses to open was Moses Haynes' saloon, and lots on the main street were selling for fifty dollars. Durfee organized the Cornucopia Mining District in August 1873, and by the end of the month there were numerous tent houses, three saloons, two restaurants, two butcher shops, a blacksmith shop, and a small school in the camp. Glowing reports published in the *Elko Independent* stirred up great excitement and brought a constant flow of people to Cornucopia. Three sections of town (Kaufmanville, Baldinsville, and Brownsville) developed, separated by only a few yards and a rise in the hill. Baldinsville was the Leopard Mining Company's company town. Brownsville was named for Henry Brown, postmaster for the Cornucopia post office that opened on November 3. Brown also served as the local Wells Fargo agent.

William Ford built a toll road from Taylor Canyon in 1873. Woodruff and Ennor purchased the road and established the Elko–Cornucopia stage line, an eight-hour trip that cost $2.50 each way. A number of stages ran, with fifty people arriving every day. Ford also built the twelve-mile Independence Valley and Cornucopia toll road, which ended at the Bull Run Road. Allen Fisher, William Haseltine, B. Davis, James Gear, and D.E. Griffith built the Deep

Remains of the mill located above Cornucopia. (Photo by Shawn Hall)

Creek toll road, while A.J. Sinclair opened the Cornucopia–Winnemucca toll road. J.W. Coffman built the Cornucopia toll road. The latter was only a mile long; it ran from Cornucopia and joined the Deep Creek road. It must have been depressing for Cornucopia residents to have to pay tolls on virtually every road going out of town.

While a number of people soon left Cornucopia after failing to get rich quickly, 1,000 residents, including 400 registered voters, crowded the town by the end of 1873. Businesses continued to open, and they included the Cornucopia Brewery, the Cornucopia Brewery Saloon, the Cornucopia Drug Store, the Cornucopia Hotel, the Cornucopia Restaurant, the Cornucopia Exchange Saloon, the City Bakery and Chop House, the Occidental Saloon and Lodging House, and the Great IXL Clothing and Dry Goods Store. A number of impressive frame buildings were completed by the end of 1873. Frank Culver built a twenty-by-thirty-foot building that housed Tim Brown's store and blacksmith shop and Mrs. Ross's restaurant. L.I. Hogel built a twenty-eight-by-forty-foot pilaster boardinghouse and saloon at the upper end of old Kaufmanville at a cost of $8,000.

But it was mining that kept Cornucopia booming. While Durfee's initial discoveries were mainly of float gold, later shafts sunk hit lode deposits. Development was slowed by a lack of suitable wood for mine timbers, the closest source of which was sixteen miles away. By the end of 1873 active mining companies in the Cornucopia District included the Leopard Mining Company (the Leopard, the Antelope, and the Constitution mines), the Plummer and Company (the Chloride mine), the Meacham and Company (the Meacham mine), the Harville and Company (the Harville mine), the Oracle Mining Company (the Oracle mine), and Kelse Austin and Company (the Kelse Austin mine). The Leopard Company, under the supervision of G. H. Willard, was the most prominent and productive. Other active mines included the Roanoke, the Constitution, the Leopard, and the Rambling Sailor. Miners earned a decent wage of four dollars a day. Most ore mined was sent to the newly opened Vance mill at Mountain City. The Leopard Company, however, shipped its ore to Giraca and Gintz's Humboldt Canal Reduction Works in Winnemucca. There was tremendous need for a local mill, because the cost of shipping ore was becoming prohibitive. The Leopard Company bought the Drew mill at Battle Mountain and moved it to a site below Cornucopia, and operations began on December 12, 1874. A new town, Milltown, formed and soon had a population of fifty.

Cornucopia's peak years fell between 1874 and 1875. The winter of 1874 was a particularly harsh one, and much of the transient population left. There was so much snow that mail service stopped, and the stages could not get through. Finally, in April, mail-hungry residents loaded outgoing mail in sleighs, transferred it to wagons at the point where the snow ended, loaded up Cornucopia's mail in Tuscarora, and made the tough trip back home.

Once the snow melted, Cornucopia's mining industry hit full stride. In July the Leopard Company reorganized and incorporated as the Leopard Gold and Silver Mining Company, with a capital stock of $3 million. The Oracle Mining Company greatly expanded its operations and began working five mines: the Oracle, the Awful Hole, the Orphan Boy, the Governor Stanford,

Mill walls at Milltown. (Photo by Shawn Hall)

Stone ruins in downtown Cornucopia. (Photo by Shawn Hall)

and the North America. Other active mines included the Grant Hussey, the Cope, the Chloride, the Black Diamond, the Constitution, and the Carrie.

Cornucopia's continued expansion included the construction of an impressive two-story, thirty-room hotel. The population had stabilized at around 500. In the fall an interesting contest took place between two men who had been pursuing the local baker's daughter. It was determined that whichever of the two won a footrace would marry her. A young man named Cherowith from Cornucopia soundly trounced his opponent, who hailed from Elko. It later came out that Cherowith had once been quite a runner, and that the nature of the deciding contest had been his idea.

The completion of the Leopard Company mill, after many delays at Milltown, brought another flood of workers to the area. The ten-stamp mill started up on December 1874, running on sagebrush because wood was so expensive. Twenty-five tons of ore arrived daily at the mill, but it could only process ten of them. Cornucopia continued to boom in 1875. New businesses included the huge Wilson and Lander stone saloon, which was 110 feet long, 34 feet wide, and 16 feet high; Pearce's Bakery and Chop House; Dickinson's Drug Store; and Sam Mooser's fireproof store. Population stood at 452, and demand was great enough for Woodruff and Ennor to make the stage a tri-weekly. The pair bought a bright red Concord stage and ran two fully stocked stores, one in Cornucopia and the other in Milltown. Though the company also ran another stage line to Carlin, its empire was collapsing. It had lost the mail contract to Smith Van Dreillen in 1874, and he took over the Elko–Tuscarora–Cornucopia–White Rock line during the summer of 1875. His stages traveled the same route in much less time, and the Carlin stage line was abandoned.

By April 1875 five Cornucopia mines were listed on the San Francisco Stock Exchange; they were the Leopard, the Cornucopia, the Consolidated, the Tiger, the Panther, and the Constitution. In May the South Leopard Gold and Silver Mining Company (M.P. Freeman, president), which owned the Elko, the New York, and the Tioga mines, was incorporated. Other new mines included the Republican, the Rambling Sailor, and the Morning Star. In April the Constitution, the Chloride, and the Tiger mines were incorporated under the name Cornucopia Consolidated. During the summer the Tip-Top mine, which the Ford Mining Company owned, began production.

Cornucopia had a problem with claim jumping throughout its relatively short life. Given that its mines were the richest in the area, the Leopard Company was normally the target. In July 1875 Samuel Hardy and Company filed suit over the Leopard Company's Champion and Miners Delight mines, claiming that Hardy had filed claims before the company did, but court papers revealed that Hardy had filed a month later, and the case was thrown out.

Disaster struck Milltown on July 24, 1875, when the mill burned after some sparks ignited a pile of sagebrush stacked next to the mill for use as fuel. The loss was estimated at over $25,000. However, the Leopard Company had purchased the old Mineral Hill twenty-stamp mill in June and was already in the process of moving it to Milltown. One hundred men worked to reconstruct the mill, which began operation in October. The first two bars poured earned the company $3,000. The mill had originally cost $140,000, but the Leopard Company bought it for only $17,500—although it did cost $30,000 to move and rebuild the mill. Despite losing one of the mills, the Leopard Company now employed fifty men and was producing $20,000 worth of gold and silver per month.

The towns of Cornucopia and Milltown continued to expand. The Dorsey brothers began operating a fast freight and passenger line from Elko. In July the Ellege and Coffman Saloon made Elko County history when it was issued the first gaming license in Elko County. The owners paid $400 to run a faro game for three months. Other recreational activities included horse racing at Milltown and dances at the Hogel building. The Hogel store was the focal point of Cornucopia. However, an invasion of crickets during the summer made it tough to enjoy the prosperous times. A justice court was established.

James Cann was appointed justice of the peace, and Thomas McAvin was constable.

The end for many mines in the district came in 1876, as veins began to fade and disappear. The large mines managed to keep going, and in June bigger hoisting works were built at the Hussey and the Leopard mines. A number of new strikes took place on the south side of Cornucopia Mountain, where the Alameda, the Boomerang, and the Lexington mines began production. On July 13 a tremendous cloudburst hit Cornucopia, damaging or destroying much of the residential area and starting an exodus from the town. On August 13 a 500-pound bucket struck popular Leopard Company chief timber man William Scott, who fell 260 feet to his death at the bottom of the shaft. He was buried in the Cornucopia Cemetery. This mining accident intensified the gloomy pall hanging over the town. At the end of 1876, the *Elko Independent* was filled with delinquent notices for assessments on Cornucopia's mines, and only the Leopard Company gave dividends to its investors, paying fifty cents a share.

The downslide continued in 1877 as mining companies folded one by one. While in 1876 the Leopard mine had produced $480,000 worth of gold and silver, the whole district earned only $133,000 in 1877. By the end of the year, population had shrunk to seventy-five, and Cornucopia's only producing mines were the Panther, the Hussey, and the Leopard. The once-mighty Leopard mine, which had produced $1 million worth of gold and silver in four years, was fading alarmingly quickly. Passenger stages were often full, but this time with people leaving. Freight traffic evaporated. Van Dreillen, who shut down the line after only a couple of months, bought the Chamberlain stage line from Winnemucca. He soon omitted Cornucopia from his other stage line's route to Elko, which now went through Deep Creek. Cornucopia's lifelines had been cut. The mill was still operating, but production continued to drop dramatically in 1878. A solid core of Cornucopians stayed on, believing a new boom would start up, but they were mistaken.

A small strike in the Leopard mine in 1879 fueled some hopes, but the mill was mainly working on old ore dumps and stockpiled ore. Most of the businesses had closed, though in 1880 a deputy sheriff and a justice of the peace remained in the town of 113. Among those left were two teachers, a grocer, a hotel keeper, a blacksmith, five prostitutes, and twenty Chinese of unstated occupations. Only the Leopard mine was producing anything, but the company that owned it was in serious financial trouble and faced a number of suits because of unpaid bills. On July 11 the mill fell victim to a suspicious fire which burned a number of homes and businesses. The company officially folded soon afterwards, and while no one was charged with starting the fire, locals had little doubt that the company was behind it. A few other mines were worked after the fire, but without much success. The Mainstay and the Ramshorn only operated for about another year before closing.

In October Wells Fargo closed its Cornucopia office because no bullion had been shipped for months. A December fire in Cornucopia spread unchecked, destroying much of the downtown area, because there were not enough men left in town to fight the blaze. Before the mill burned, sixty people still lived in Cornucopia and Milltown, but by the end of 1880 only about twenty were left. The school was still open at Milltown, but closed early in 1881 when there were no more students. Cornucopia died completely in 1881. In February the IOOF Lodge no. 29, which had been meeting in Hogel's hall since 1875, moved to Tuscarora. Billy Hoar closed his hotel, the pride of Cornucopia. By March, Hogel's was the only business left in town. In a show of bravado, Hogel made plans to build an opera house, and a cornerstone was laid in April, but construction never started.

A small strike at Good Hope, located to the west of Cornucopia, led to the establishment of two stage lines. Cluggage and Parker Stagelines ran the U.S. stage from Battle Mountain to Cornucopia via Tuscarora, and J.B. Ringgold ran one to Good Hope. Ringgold was also owner of the Cornucopia Hotel, which he reopened for a time while running the stage. Ore from the limited mining at Cornucopia was shipped to the Good Hope mill. In June the Leopard Company sold its tailings for $1,500 at a sheriff's sale to settle a judgment for the Owyhee Land Association, and by November all mines had been abandoned.

Whitney, Vessey, and Company began construction on a mill to treat the tailings in October 1881. The company contracted with Lew Willard to haul 200,000 pounds of equipment, and the mill, built next to the Leopard mine's dumps, was completed in the spring of 1882. By July the facility, under the guidance of M.J. Hatch, was producing $2,500 worth of bullion a week, but the company ran into financial troubles, and creditors took it over in October. Lew Willard reopened the mill in March. Edgar Reinhart and Company purchased the mill in April, put it up for sale in June, and shut it down for good in August. In September 1883 the mill was dismantled and moved to Tuscarora, where it was sold and shipped elsewhere.

By the fall of 1882 Cornucopia's two remaining stage lines had stopped running. In November Antone Storff of White Rock bought the Greenbaum and Greatzer building and used the material to build a granary on his ranch. Many other buildings were also sold and removed. By the end of 1882 only three residents were left in Cornucopia. Some mines operated sporadically during the 1880s. They were the Pride of Nevada, the Monarch, the Big Horn, and the East Leopard. However, nothing rich enough to draw people back to Cornucopia turned up. The post office, which had not seen much business for years, finally closed on October 16, 1883, when postmaster Owen Vaughn left town. This left Thomas McAvin as the only resident. He gave up the ghost in early 1884 and went to Tuscarora, where he opened a business. The once-mighty town of Cornucopia was no more.

Some mining activity has occurred in the area over the past hundred years, but Cornucopia never revived. In 1890 Seth Roseberry and James Baker leased the Leopard mine and removed some low-grade ore. Will and Seth Roseberry bonded all the Cornucopia mines in 1897, but the mines produced little. By 1908 all the deep workings in Cornucopia's mines had collapsed. Hopes for a revival were rekindled in 1915 when George Russell reopened the Leopard, but they fizzled out before any production occurred. In 1918 Ed Peacock found some paying silver ore on his claims and made a small profit before giving up in 1919. The only significant production occurred from 1937 to 1940, when the Pan Mining Company began reworking dumps of the Cornucopia and the Leopard mines. N. W. Parker, Martin Arrestoy, and K. L. Reed owned the company, and tailings were sent to the McGill smelter. In three years they recovered 1,500 ounces of gold and 38,000 ounces of silver.

In March 1940 the La Plata Mining Company, which James Truitt of Tuscarora owned, consolidated the old New York claim group, which consisted of the Hussey, the Panther, the Mohawk, the Constitution, and the Republic mines. The company re-timbered the Leopard mine and planned construction of a 100-ton mill, but the plan never materialized and Truitt gave up within a year. There was no more activity in the district until 1974. Spartan Exploration, Limited, did some exploratory drilling, as did Homestake Mining Company in 1979 and Western States Minerals in 1981. A small open pit leaching operation was active from 1985 to 1989 but met with only limited success. Total production for the Cornucopia District is valued at well over $1 million. Production that took place from 1872 to 1882 was responsible for all but $110,000 of that amount.

The remains at Cornucopia today do not begin to hint at how active the town once was. At the townsite, a couple of stone cabins are left, along with the stone walls and foundations of many other buildings. Up on a rise beyond the stone cabins, more stone walls mark the site of old Kaufmanville. Just above town, the remains of the open pit operation show where the Leopard mine once stood. Surprisingly, the old boiler is still within stone walls, surrounded by mounds of processed ore—the only sign of the glory days of Cornucopia's mining. The Cornucopia Cemetery, which had a number of graves, has virtually disappeared. Many graves were moved after Cornucopia collapsed in the 1880s, and only two stones remain. The cemetery is now in the midst of heavy sagebrush, and only mounds reveal its location. At Milltown, the ruins of the two mills dominate the townsite. The large walls of the twenty-stamp mill dwarf the remains of the ten-stamp mill. Numerous stone ruins and foundations are below the mill ruins.

Members of the Oldham family in front of Dinner Station, 1907. (Tony and Ellen Primeaux collection, Northeastern Nevada Museum)

Dinner Station

(Coryell's) (Dorsey's) (Weiland's)
(Winters') (Oldham's) (Park's)

DIRECTIONS: *Located 23 miles north of Elko on Nevada 225.*

Dinner Station, known by many names during its history, was the most prominent stop on the Elko to Tuscarora stage road. The station also served stage lines heading up to Mountain City. The first station was a wood-frame building that Alex Coryell built in the late 1860s to serve as a stop on Hill Beachey's Elko–Idaho route. The property sold for $12.70 in May 1870 at a sheriff's sale, and John Dorsey took it over. During the 1870s a constant stream of traffic, including Van Dreillen's Elko–Tuscarora–Cornucopia–White Rock stage, flowed through the station. The fare to Dinner Station from Elko was three dollars. In addition, Dinner Station was the terminus of the North Fork–Gold Creek Stage Company, a tri-weekly stage run by Will Martin.

J. H. Weiland, who ran an unofficial post office, distributing mail locally and making sure it was transferred to the proper stages, bought the station complex in 1880. In 1884 the station burned to the ground, but by October Weiland had completed the building of a new two-story stone station that was "the handsomest and most comfortable wayside hostelry in the state of Nevada."[1] Fred Wilson, a freight-line operator from Tuscarora, sold subscriptions to help Weiland raise money to rebuild the station. The new station had a "kitchen, dining room, living room, and two bedrooms on one side. There were three double bedrooms and two single rooms for guests on the side nearest the road."[2] A controversy developed between Weiland and the Elko–Tuscarora stage line in October 1886. The stage line had ordered Weiland to pay fifty cents per head in addition to providing free board for stage employees. In protest, Weiland stopped serving dinner to passengers, and the stage line made Reed Station the dinner stop. Within a couple of months, Weiland and the stage line struck a compromise, and Weiland began providing full service by the end of 1886.

In July 1888 Weiland opened a saloon and small store adjacent to the station house. A school opened in 1889. Fannie Grant was the teacher. The station house fell victim to fire once again in September 1890, and Tuscarora residents raised more than $500 to help Weiland refurbish the station. Because the stone structure was undamaged, reconstruction was quickly completed. While the station was closed, operations ran out of other buildings at Dinner Station. There is an interesting story about foreigners visiting Dinner Station for a meal. It goes as follows. "It was the custom during summer for the plates to be set out on the table upside down so flies could not walk on the side where food was to be served. The serving dishes were passed and the foreigners piled the food onto their plates which were still upside down. They apparently believed that upside down plates were a western table custom."[3]

Weiland fell ill in May 1900 and died of the grippe. He was born in Pittsburgh, Pennsylvania, had first come to Elko in 1876, and served in the Nevada Assembly in 1894. He left a wife, four sons, and four daughters. His son, Alex, took over the station. At this time, the census showed that forty people lived around the station. Mrs. Weiland sold it to Frank Winters in October. The new boom at Jarbidge led to an increase in traffic, and Winters hired Chinese cooks to handle the additional work. Winters had a reputation for helping unfortunate travelers. Walt Davidson, a Swede, was on his way to Mountain City from Minnesota to join his brother, Jack, and he ran out of money by the time he boarded the stage at Elko. All he had left to eat was cheese and crackers. During the stopover at Dinner Station he went off behind the corrals to consume his meager lunch. Winters saw this and invited him inside to eat. He asked Walt if he was Jack's brother, and upon hearing that he was, Winters shook his hand, squeezing a five dollar bill into it and saying, "you can pay me back when you have it."[4] Davidson never forgot this act of generosity and always cried when he told the story.

Ed Oldham bought Dinner Station in 1905, and as many as seventy-five travelers a day ate lunch there. The station could sleep twenty, and the barn held seventy-five horses. Oldham charged fifty cents each for meals and lodging and continued to run the store and saloon. The Oldham family sold Dinner Station to Tom Parks and moved to Fox Springs, also a stop on the Elko–Tuscarora stage line.

Over the years, a good-sized ranch had grown up around Dinner Station. With the advent of the automobile, need for the station diminished. In 1915 Joe and Frank Yraguen bought the ranch and station. From 1918 to 1924 Gertrude and Tom Eager managed the complex. They also ran a stage line from Dinner Station to Gold Creek carrying passengers, mail, and freight. Ninety horses were housed in the big barn. The school at Dinner Station was active into the 1920s.

The Moffat Company bought the station in the 1940s, and E. L. Cord purchased it in 1960. David and Marion Secrist bought it in 1972 and completely renovated the station house. Frank and Phyllis Hooper, the current owners, later bought the station. Unfortunately, in October 1991 a fire once again caused the station extensive damage. Luckily the Hoopers, who could have taken the cheaper route and razed the structure, decided to invest a great deal of time and money into restoring the building. The final touches were completed in 1994.

Divide

DIRECTIONS: *From Tuscarora, head south for 1½ miles. Turn right and follow the road, keeping left, for 7 miles. At the fork, head right and follow the road for 5 miles to Divide.*

Divide, so named because it was on the divide between the Humboldt and Snake River basins, came into being in 1915. W.C. Davis and J.L. Workman, who were instrumental in the early discoveries at National (located in Humboldt County), discovered the San Juan mine in July. A small rush started from Tuscarora, and within weeks more than fifty miners had arrived in the tent camp.

By August Divide's population was seventy-five. The Divide Mining District was organized on August 31 (William Clark, recorder). Barney Horn set up the town's first business—a saloon and general store—in a twenty-seven-by-thirty-eight-foot tent. During the next year little was produced, but a considerable amount of development work took place. Ore was being shipped through Red House, and R. W. Bender alone had twenty men working on his claims. Three wood buildings had been built and three more were nearly complete, but the ore veins were short, and by 1918 the camp was empty

Mine shaft at Divide. (Photo by Shawn Hall)

Only building left at Divide. (Photo by Shawn Hall)

except for a couple of prospectors. The district produced only a little more than 1,300 ounces of silver.

It was not until August 1928 that mining resumed. A silver and gold vein in the April Fool mine, which Jack Walter, Kenneth Dale, W. P. Young, and Mrs. L. Pendergast owned, proved valuable. The group hired Kirk Cornwall to haul ore to the Gold Circle Consolidated mill in Midas, and in September sold the mine to A. Backlund for $65,000. However, Backlund quickly discovered just how shallow the Divide ore veins were, and he gave up within a few months, much wiser and poorer for the experience. Since 1928 no mining has taken place at Divide. Some exploratory drilling has been done over the last ten years, but a substantial ore body has not been located yet. One building, a collapsed cabin, and a couple of shallow shafts mark the site of the short-lived town of Divide.

Old Heart of Nevada

Edgemont

DIRECTIONS: *From White Rock, head south on Nevada 226 for 4 miles to Edgemont.*

While the town of Edgemont did not form until 1900, mining had been taking place along White Rock Creek, just to the north, since the early 1870s. This activity led to the establishment of the town of White Rock City. During the 1870s the main mines in White Rock Canyon were the Central Pacific, the Porter, and the Town Treasure. Pack mules hauled ore to the Cope District mills. J. A. Savage, W. D. Porterfield, and Virgil Bartlett planned a mill for White Rock Ravine, but they never built it. The mines were abandoned by 1880, and little mining was done in the area until 1897. At the time, most mining was occurring on the east side of the Bull Run Mountains at Blue Jacket, Bull Run, and Columbia. In 1897 the White Rock Gold Mining Company was organized. This placer operation encompassed sixteen claims along Silver Creek in Borette Canyon. The best claims were the Palo Alto, the Matchless, the Surprise, and the Climax. The company spent $38,500 on preparatory work, which included the construction of a 400-foot flume and a three-quarter-mile water ditch. By July 1897 the placers were finally being worked, and production averaged about four dollars per cubic yard.

The discovery in 1898 of the Lucky Girl mine, located 1,200 feet above the town, led to the beginnings of Edgemont. Alex Burrell, who organized the Montana Limited Mining Company in May 1901, bought the Lucky Girl and other nearby claims from the Walker brothers. They immediately put into motion plans to build a twenty-stamp mill at Edgemont and created

Overview of Edgemont, 1906. (Reginald Coffin collection, Northeastern Nevada Museum)

a company town during the summer. Parts for the new mill started to arrive, with much of the machinery coming from the Drum Lummond mine in Marysville, California. In September four seven-ton iron battery blocks arrived, but they were too heavy to haul to the mill site and were sent off to be refitted with wood. A post office opened on October 1, 1901. Hubert Holt was postmaster. A population of close to 150 now crowded into the narrow canyon. The Northern Stage Company's western branch added the town as a stop, with a fare to Edgemont of five dollars from Tuscarora and eleven dollars from Elko. The twenty-stamp Lucky Girl mill started up in January 1902, with a tramway bringing ore from the mine. The mill processed sixty tons a day, employed seventy-five men, and ran on electricity from a power plant located near White Rock. The Lucky Girl mine hit its stride in 1903 and led Elko County gold production for the next four years, which helped revive the county's sagging mining industry.

The town continued to grow. William Ennes Clawson and his wife, Agnes Thompson, moved to Edgemont and built a three-room log cabin. Their son, Marion, who was born in Edgemont, became director of the Bureau of Land Management in 1948. Agnes's sister, Jennie, with her husband, Al Johnson, built and ran the Johnson Hotel, boardinghouse, and restaurant. The two families later moved to Bull Run during that camp's short-lived boom. A school district was established in September 1902, and Celia McCarty was the first teacher. William Hanson, who formerly ran the Baldwin Store in White

Rock, bought the Currey Saloon. Currey turned his interests to building a hall, which served as the Edgemont community center.

A problem that continued to plague Edgemont throughout its history first reared its head in February 1904, when an avalanche carried away the hoisting works at the mine and damaged the mill. All the miners and mill men were released until repairs were completed in April. In May Paul Lovelace and James Ross, disgruntled former employees, were caught a few miles out of town after they had stolen $2,500 in amalgam from the newly refurbished mill. By 1905 the Lucky Girl mine was the largest deep mine in Elko County.

Another avalanche in January 1906 destroyed most of the wooden shell of the mill. While the stamps were undamaged, the avalanche swept away an undetermined but large amount of bullion that was lying on the mill's plates. The mill reopened and was running by the end of February, and by summer the Montana Company employed ninety men. In 1907 the company added a sixty-ton cyanide plant, run by electricity, next to the stamp mill.

The Lucky Girl was again the largest producer of gold in Elko County, but dark clouds loomed on the horizon. The ownership of the Lucky Girl mine was challenged in court, and the mine and mill closed in 1909. From 1902 to 1908 it had produced more than $745,000 in gold and silver. In 1910 Edgemont's population was still 150, but everyone was holding on by a thread, because they harbored hopes of a quick legal resolution. These hopes were dashed as the years passed. By 1915 Edgemont was a virtual ghost town, and the school moved to the Mitchum Ranch in the valley below.

Another avalanche in 1917 heavily damaged the idle mill and destroyed many of Edgemont's buildings. The post office, normally a town's last sign of life, closed on October 15, 1918. C. C. Randall, J. F. Worman, George McGuire, and Harry Sears took over the mining properties. In May 1932 Atlas Gold Mines bought the group out and put H. H. Carpenter in charge of reopening the mines and mill. A fifty-ton Marcy ball mill partly replaced what was left of the old mill. Production did not commence again until 1934 and never came close to the figures reached in the early years.

Other companies made a try in the district during the revival. In May 1937 the Echo Canyon Mining Company was organized and worked seven claims, which the Boyce brothers originally located in 1925. The company built a 600-foot tram to their twenty-five-ton concentration mill, which opened in July 1938. The company enlarged the mill to forty tons in September and built a cookhouse, a bunkhouse, and cabins on the property, located just north of Edgemont. In November 1937 George Boyce, president, formed the White Rock Gold Mines to work claims to the south of Edgemont, and the British Columbian Company, the Bralorne Mines Corporation, reopened the Edgemont mine in May 1938. The companies worked the mines until 1942. During this period, the mines employed about thirty men and produced only 16,000 ounces of gold and 3,000 ounces of silver. Lead was the major mineral mined, for a yield of 200,000 pounds. After 1942 only sporadic, limited production took place. The total value of gold, silver, and lead that Edgemont's mines produced is just under $1 million. The present owner bulldozed remaining buildings because of liability worries, and only mine dumps and mill foundations are left. The site is located on private property.

Falcon
(Rock Creek)

DIRECTIONS: *From Divide, continue for 3½ miles, then turn left and follow the road for 1½ miles to Falcon.*

James Dunphy settled on Rock Creek in 1870, and the Dunphy Ranch, located near Beowawe, used the Rock Creek Ranch as a base camp. John Epley moved to the Rock Creek area from Lamoille in 1873 and prospected the canyon with little success. Silver was discovered along the creek in July 1876, and on August 24 the Rock Creek Mining District was organized. The Falcon Mining Company was incorporated on September 10 with M. P. Freeman, prominent Elko resident, serving as president. The company operated three mines: the Falcon, the Eagle, and the Manhattan. The Falcon mine, under the supervision of R. D. Norton, was the main producer. Other mines which started up included the Scorpion, the Lookout, the Challenge, the Aspin, the Hunter, and the Flagstaff.

Not much happened at Falcon in the winter, but the level of activity greatly increased in 1877 and 1878. During the summer of 1877 a small company camp named Falcon formed. At the same time, Ed Banning extended his freight line from Tuscarora to Falcon on a road Len Wines built. In November Jack Gaston started a daily stage to Tuscarora. The Tuscarora and Rock Creek toll road opened, and the Wines brothers organized the Tuscarora and Rock Creek Timber and Tollroad Company to collect money from people cutting timber around Rock Creek.

In October a San Francisco syndicate bought a half interest in the Falcon mine for $25,000, and ten tons of test ore processed at the Leopard mill in Cornucopia returned $166 per ton. Soon afterwards, the mine was listed on the San Francisco Stock Exchange. The Scorpion mine became active that same month, with initial assays running as high as $550 per ton. The Scorpion Mining Company was formed but was renamed the Rock Creek Mining Company and incorporated the following month. R. D. Norton, striking out on his own, bonded four new discoveries—the Silver Belle, the Silver King, the Raven, and the Great Divide mines.

Falcon continued to grow. James Mahan, William Logar, and Elko County surveyor, R. M. Catlin, surveyed the Battle Mountain, Squaw Valley, and Rock

Creek toll road, which joined the main Battle Mountain road seven miles from Rock Creek. The Truckee Lumber Company built the road, and Carliss, Hamilton, and Company ran it. By the end of 1877 Falcon's population stood at eighty, and the camp contained twenty houses and thirteen tents. Fred Stofiel and J.D. Hoover ran four businesses: a two-story hotel called Falcon House, a store, a restaurant, and a saloon. Two other saloons also opened, and in December the first family moved to Falcon.

By 1878 Falcon seemed destined to become the latest boomtown in Elko County. On January 15 a post office opened, with Walter Stofiel as postmaster. On February 16 Abe Mooser and John Mayhugh incorporated a new mine, the What Not. Two other mines, the Ella and the Virginia, opened, and by spring many people were building new homes. But in April water started seeping into the Falcon mine, the mainstay of the camp. The water continued to rise gradually despite attempts to stem the flow. Around the same time, the first and only killing took place at Falcon, when a bar fight at the Hoover and Stofiel saloon escalated, and a man named Read knifed another man named Bohn. Bohn pulled out his gun and shot Read in the head. The death was ruled justifiable homicide.

Stockholders focused on the Falcon mine's problems, and during the summer all work at the mine suddenly ceased. As a result, the post office closed on August 27. On September 26 the mine, machinery, office, and tramway brought only $248 at a sheriff's sale held to satisfy a judgment for Jon Schuck of Rock Creek. After the sale, assessment after assessment took place, but despite promises of renewed activity the mine remained idle. Most of the

The Falcon Mine today.
(Photo by Shawn Hall)

town's residents left, and by winter many businesses had closed and all stage and freight lines had stopped operating. Little occurred in 1879, although some small shipments of ore were made to mills in Tuscarora. In April John Epley, who decided to return to Falcon after many years, sold the Oriole, the Arcade, the Bobby Burns, and the Cargo claims to J. G. Berry. The Falcon mining property was sold at a sheriff's auction in November for $3,720.

By the summer of 1880 only the Bajazette mine, which Epley, who was now district recorder, owned, was active. In October Epley re-timbered the Falcon mine down to the 225-foot level, but the other two levels were still under water. He opened a new mine, the Longfellow. The Independence mill in Tuscarora processed three bars of bullion that returned $4,038, but the tough rock made progress in the mine very difficult. Epley reorganized the Falcon Mining Company in 1881, and under his guidance the company soon controlled twelve mines, including the past main producers, the Falcon and the Scorpion. The renewed activity breathed life back into Falcon, and a voting precinct was established with William Roberts as voter registry agent.

In October construction began on a mill, and Charley Warner hauled 50,000 feet of lumber from Elko. Mill contractors Walter, Lind, and Company of San Francisco arrived to begin construction, but a foot of snow fell in early November and prevented the mill machinery, already en route, from reaching Falcon until December. The mill was completed during the summer of 1882, but it was a wasted venture because, while the ore was rich in silver, it was too hard, and no one could come up with a method by which to successfully process it. The mill never yielded one bar of silver. It was dismantled in 1884, although the machinery stayed behind. In August 1887 F. F. Coffin purchased the equipment for use on the Nevada Queen and Commonwealth properties in Tuscarora. A last try took place in 1888, when the Silver Chariot (formerly the Scorpion), the Casey, the Lizzie, the Moscow, the Morning Star, and the Diamond (the last three of which were worked by George Jenkins, Miles Dunton, William Jewell, and Fred Wilson), were worked for a short while. The dream that had been Falcon was dead.

A small revival of some of the mines took place in the 1920s. The Ruby King produced some silver during the early part of that decade, and the April Fool group, which Walker and Young owned and which included the old Falcon mine, produced 20,000 ounces of silver in 1928 and 1929. In 1937 Tom Wylie and Walter Maxwell reopened the Falcon mine, and from 1937 to 1942 they mined 3,000 ounces of silver before giving up.

Since that time, not much more than occasional exploration work has taken place at Falcon. A number of cabins still stand there, and old mine dumps mark the core of the main mines. The Falcon mine has mining shacks and equipment left from the activity of the 1930s and 1940s, while a couple of stone ruins and foundations mark the old Falcon townsite.

Good Hope
(Aurora) (Amazon) (Walker City)

DIRECTIONS: *From Jack Creek, head north on Nevada 226 for 6 miles. Exit left and follow for 15 miles. Turn left and follow Chino Creek for 5 miles to Good Hope.*

Initial silver discoveries at Good Hope occurred in 1873. During the summer the Amazon Mining Company and the Silver Brick Mining Company began small-scale production. Both companies worked mines along Amazon Creek, and the Amazon Mining District was organized in the fall. In 1875 the owners of new mines along Chino Creek, about a mile west of Amazon Creek, organized a new mining district called Aurora. Its four prominent mines were the Buckeye, the Ohio, the Page and Kelley, and the Snyder. To confuse matters, the Walker Mining District was also organized just to the north, and the camp that had formed along Chino—now called Four Mile Creek—was named Walker City. It consisted of ten miners' cabins. This confusion was resolved when the Good Hope Mining District, which encompassed the other three, formed in January 1879.

By March 1879 the district's principal active mines included the Silver Brick, the Buckeye, the Aurora, the Snyder, the Ohio, the You and I, the Page and Kelley, the Atlantic, the Cable, the Rattler, the Trade Dollar, the Arizona, the Ella, the Amazon, and the Tiger. Timbers for the mines came from the Bull

Ruins of the Good Hope Mill. (Photo by Shawn Hall)

Run Mountains and were hauled to Good Hope, at considerable expense. In May the Aurora sold for $30,000, but the new owners struck water the following month and the mine closed. The Amazon mine met with the same fate in September but was bonded to Tuscarora parties for $20,000.

By 1880 Good Hope's population was thirty-six, and the ore found in the district was deemed sufficiently valuable to warrant the construction of a mill. High costs of ore transportation to Tuscarora threatened the future of Good Hope, and construction began on a five-stamp mill in November. The cost of shipping the mill machinery was incredibly high. Parts were scavenged from other relatively nearby mining towns. These included a boiler from the Grand Prize mine in Tuscarora, which was converted into a roaster at the Dove and Jack Foundry, and an old engine brought from the Leopard mine in Cornucopia. H. W. Taylor and Charles Russell built the mill using machinery that part owner James Prout and James Dove installed. The mill was initially called the Ballantine, after part owner Tom Ballantine. It was fired on May 23, 1881, and it worked a load of Tiger mine ore.

Hope became a bustling mining camp. A number of businesses opened, including two saloons, which Mark Flemming and Harry Wheeler owned. E. O. Connors built a boardinghouse and also furnished the mill with brush and hauled ore. Two stage lines began running to Good Hope. The North Fork–Gold Creek Stage Company ran a tri-weekly stage from Elko with a fare of nine dollars. J. B. Ringgold ran a stage from Cornucopia. Besides passengers and freight, he also hauled ore from the Cornucopia mine dumps for processing at the Ballantine mill. In July Ballantine sold his interest in the mill, and the owners, Primeaux, Cockbin, and Prout, renamed it Good Hope. L.E. Atchison's Amazon mine was now a 100-foot shaft producing ore worth $300 a ton. Atchinson moved one of the abandoned buildings at Cornucopia to the Amazon mine to provide housing for the miners, but he left Good Hope in September to take charge of a mill at Phillipsburg, Montana. In the same month the Amazon Mining Company was incorporated, with L.I. Hogel as president and treasurer and W.W. Rogers as secretary. In October the Buckeye and the Ohio mines consolidated, forming the Consolidated Ohio Mining Company, and the mine became the most prosperous in the Good Hope District. The ore in the district proved rich but tended to occur in small pockets, and by the end of 1881 a number of the early mines had played out. Besides the Buckeye and Ohio and the Amazon, other active mines were the Snyder, the Ella, and the Silver Brick. The Buckeye and Ohio continued to be Good Hope's main producer and kept the mill running around the clock. In 1883 many new mines opened, including the Patience, the Tiger, the Tough Nut, the Barrus, the Imperial, the Golden Era, and the Midnight Star. Hain also filed for a mill site along Amazon Creek, but the mill was not built. Charlie Grove, Jim Wilson, and Newton Ballantine bought the Amazon mine and renamed it the Big Muddy. About twenty men worked in the district.

In January 1884 the Ohio Consolidated Company expanded its operation and hired twenty more men. A post office opened on March 24, with William Pool as postmaster, and mail came from Tuscarora once a week. But the bottom fell out of Good Hope during the fall of 1884, when the Ohio Consolidated Company left the district. From 1882 to 1884 $100,000 had been taken out of the Buckeye and Ohio. The mill also closed down, as did most of the other mines, and by the end of the year all stage lines had stopped running to Good Hope. One of the camp's residents had to pick up mail in Tuscarora. In September 1885 Dave Hain reopened the Buckeye and Ohio, but most of the rich ore pockets had already been mined out. Only a handful of miners remained, searching for the elusive ore pockets. The Gerald, the Langley, the Tioga, the Lewis, the Let Go, the Big Gem, and the Langtry were still being worked, but they produced little. The post office finally closed on February 3, 1887, because of lack of business. By 1888 only the Thisle, the Hudson, the Protection, and the Blue Ruin were being worked, and by 1890 all of them had been abandoned.

It was not until 1902 that activity started up again when C.W. Jordan of Omaha, Nebraska, bought the Smuggler mine in February. He organized the Haggerty-Jordan Copper Mining Company to work the mine. In addition, the Iowa Chief Mining Company developed several claims and installed a gasoline hoist at their new Aspen mine. A.M. Miller managed the mine and Lee Faison, one of Good Hope's old timers, was part owner. The following year, the company built a fifty-ton concentrating mill, but it was unsuccessful and closed by the end of summer. The company folded soon afterwards.

Mining returned to Good Hope in 1918 as a result of a rich strike in the Midnight mine (formerly the Buckeye and Ohio), which R.M. Lesher and J.W. Reed owned. Charles Woodward, who was leasing the mine, made the strike. By 1920 Good Hope showed signs of reviving. Irwin Chandler and R.K. Humphries took over the Midnight mine and extended the shaft to 700 feet. The Snyder mine also reopened, and Lesher, who had recently inherited $5 million, spent $45,000 refurbishing the old mill and adding a flotation system. Some shipments of silver ore took place during the next couple of years, but the revival ended in July 1923 when lightning hit the mill and fire destroyed it. Lesher sold his remaining ten claims to the Randsburg Company and left to spend his fortune elsewhere.

Good Hope's last bout of production took place in the early 1950s, when the Buckeye and Ohio produced several thousand pounds of antimony. But since then there have only been some minor exploration attempts in the area. Today, the old camp of Good Hope is marked by the foundations of the two mills, located side by side. A couple of stone foundations remain at the townsite. The Buckeye and Ohio, located beyond the mills, contains the best ruins at Good Hope, and some buildings are left below the extensive mine dumps. A gallows frame marks the Good Hope mine, which was never actually a producer, and many other mine dumps are scattered along Four-Mile Creek. Along Amazon Creek, mine dumps and dwelling depressions are all that are left of the 1870s camp of Aurora, or Amazon.

Ivanhoe

DIRECTIONS: *From Midas, head east for 11½ miles. Turn right and
follow the road for 5 miles to Ivanhoe.*

Before 1900 cinnabar ore was discovered at Ivanhoe and miners
filed 125 claims, but no further development took place. W. F. Roseberry and
W. C. Davis developed the first mercury deposits in the area in 1915. Mercury
was a rare commodity in northeastern Nevada. Roseberry and Davis organized the Ivanhoe Springs Mining Company and built a six-pipe Rocca retort
furnace. They added another furnace in 1916, and the company produced
fifty flasks of mercury before abandoning the mines in 1918. In December 1920 Roy and Elmer Roseberry bought the Ivanhoe cinnabar mines for
$6,038 at a sheriff's sale, but the new owners did little work on the mines.

It was not until July 1927, when Frank Bowers, Joe Wilson, William Engle-
bright, and T.T. Fairchild filed numerous claims in the district, that interest
in mining the Ivanhoe area revived. Three mines, the Butte, the Governor,
and the Silver Cloud, were developed. The Butte Quicksilver Mining Company ran the Butte, and Richard Estridge of Winnemucca managed it. Former
Massachusetts Governor James Curley, who organized the Governor Mercury

Mining Company with partners L.V. Pangburn and Waddy Hunt, owned the Governor. In 1930 the Governor mine produced 107 flasks of mercury. The New Verde Mines Company worked the Silver Cloud mine, which J.T. Miaddeford owned.

Andrews built a twenty-ton rotary furnace in 1930, but he died soon afterwards and the furnace was not started up until 1931. Curley built another furnace at the Governor mine, and the Mayfield Mining Company, which was working the Frank Bowers claims, completed a retort furnace on its property in May 1930. During the 1930s the area sustained a small but steady rate of production. Waddy Hunt, still part owner of the Governor mine, began working the Coleman claim group in the mid-1930s. In December 1939 he sold the claims to his friend James Curley. Curley formed the Curley Luck Gold Corporation and built a twelve-ton furnace on the property, but the venture proved unprofitable. He then concentrated his efforts on the Governor mine and built a 100-ton rotary furnace, which he started on the property in September 1940. This made 1939, when Ivanhoe produced mercury valued at $12,000, the town's best year. However, World War II caused a mining slowdown, and by 1944 all operations had ceased. From 1927 to 1944 Ivanhoe produced 2,136 flasks of mercury, 1,032 of which came from the Butte mine.

After 1944 Ivanhoe produced only small amounts of mercury. During the mid-1950s active mines were the Staggs and the Quilici, and the Clementine mine saw minor production from 1959 to 1961. In 1960 the Bogdanich Development Company mined a small amount of uranium, and during the 1960s the Fox and Sheep Corral mines had limited production. Beginning in 1980, with the emergence of microscopic gold mining, the Ivanhoe District became the scene of intense drilling. Homestake, Placer Amex, Kennecott, and U.S. Steel (now known as USX Corporation), all tested the district, and by 1984 USX had outlined a large gold reserve that Cornucopia Resources purchased in 1987. Galactic Resources joined Cornucopia, and two of the companies' subsidiaries, Ivanhoe Gold Company and Touchstone Resources, began working the Hollister deposit as an open pit, heap-leach operation. They poured the first gold in October 1990. In 1992 Newmont Mining Corporation purchased Galactic's interest and half of Touchstone's holdings. At this time, however, the Hollister mine had been virtually mined out after producing approximately 500,000 ounces of gold. The joint venture is still leaching the remainder of the ore and conducting exploration throughout the area. Most of the older mercury mines have been obliterated by the Hollister open pit and newer exploration. There are mine dumps and roaster ruins at the Governor (three miles west of the Hollister pit), the Colemen (three miles west of the pit), and the Silver Cloud (five miles southwest of the pit). Dumps and workings remain at the Fox and Sheep Corral mines.

Jack Creek

(Anderson) (Jackson)

DIRECTIONS: *Located 15½ miles north of Taylors on Nevada 226.*

Jack Creek was named in honor of its original settler, Jack Harrington, who homesteaded in 1868 and spent the rest of his life ranching at Jack Creek. Jack Creek soon became a stop on the Northern Stage Company's line from Tuscarora to Mountain City, with a fare of two dollars to Jack Creek from Tuscarora. The Jack Creek area became the prime source of wood for Tuscarora, and both firewood and mine timbers came from there. From 1877 to 1892 harvesting averaged 12,000 cords of firewood and 200,000 linear feet of mine timber. There was limited mining interest in the Jack Creek area in the 1870s after Chesley Woodward, a local rancher who had settled at Jack Creek in 1869, staked a number of claims and formed the Woodward Mining District in May 1877. However, nothing was ever produced. Woodward abandoned the claims in 1878, although he and his family operated a ranch, a store, a restaurant, and a rooming house at Jack Creek for many years. In June 1879 the Jack Creek School District was organized, and Nevada Hardesty Griswold, then seventeen, taught ten students. A settlement of about twenty people had formed at Jack Creek, and a few other ranches were homesteaded in Jack Creek Canyon and nearby areas to the north and south. For many years, there were two schools in the Jack Creek area, each with about ten students.

Because of the number of families in the area and the absence of a proper gathering place, Harrington built the Jack Creek Opera House, which he completed in November 1880. While it had an elegant moniker, the building was more or less a large barn with a stage, but the local residents enjoyed the entertainment and camaraderie they shared there. In 1884 the town had grown enough to name Harrington justice of the peace. Clay Hardesty was the constable and R. D. Lamham the road superintendent. A voting precinct was established, and local resident Charles Woodward served as the voter registry agent. Jack Harrington died in 1886, and Frank Culver purchased his ranch and stage stop in April. Culver built a large lodge for travelers and fishing parties, promising that "clean beds and good meals will be the rule."[5] Many citizens from Tuscarora and Elko came to Jack Creek to spend weekends fishing and hunting.

In May 1889 the Jack Creek School District Number 22 was abolished, and the Jackson School District Number 40 formed. There were no longer two schools at Jack Creek. The residents turned the abandoned Jack Creek school into the local dance hall. In 1890 the townspeople applied for a post office named Jackson, and on June 25 the post office, with William Clawson as postmaster, was established. But the government rescinded the postal establishment order on March 10, 1891. A number of postal histories erroneously place the Jackson post office in the eastern part of the county at the Jackson mine, north of Tecoma. However, according to Clawson's son, William E. Clawson Jr., his father resided at Jack Creek after selling his ranch in Independence Valley. In fact, the *Tuscarora Times-Review* reported the birth of William Clawson's son at Jack Creek in November 1890.

In December 1897 a fire started by sparks from the wood stove seriously damaged the schoolhouse, and the children had an extended Christmas holiday while the school was being repaired. In August 1898 the Tuscarora-based Dexter Mining Company began construction of a large power plant on Jack Creek. The old Defrees mill in Taylor Canyon was dismantled for its wood to be used in building a boardinghouse and bunkhouse at the power plant construction site. Power lines ran to the forty-stamp Dexter mill in Tuscarora, and the plant began producing electricity in January 1899. Heavy snowfall seriously hampered the final phases of construction. The water pipeline, buried as deep as ten feet beneath the snow, was 3,000 feet long. A flume brought water from Jack Creek to Chicken Creek, and the water then entered the pipe and dropped 400 feet to the plant. The plant generated more than 900 horsepower, and the power system cost $50,000. The plant brought many new residents to Jack Creek. A new store, which Christian Anderson owned, opened in August. By 1900 the population of the Jack Creek area was seventy-four. On April 2 a post office named Anderson opened. Anderson, the postmaster, operated out of his store. Anderson also owned Taylor Station, located in Taylor Canyon near Tuscarora.

J. R. Plunkett, a relative of the *Tuscarora Times-Review* publisher W. D. Plunkett, was seriously hurt at Jack Creek while loading sheep in June 1903. His horse became startled and threw him under a moving wagon. While he suffered no broken bones, the paper reported that he was not at all happy with his horse.

The Anderson post office closed on September 30, 1905. Jack Creek's population shrank drastically once the power plant was operating efficiently, and most workers moved back to Tuscarora. The power plant was completely refitted and enlarged in 1910, but in June 1911 it was struck by lightning. Initial newspaper reports declared that the plant had been completely destroyed by fire, but a few days later it was already running again, and the same sources sheepishly admitted that only the transformers had burnt out. Gradually, Basques began to purchase ranches in and around Jack Creek, and by the 1920s it had become a small Basque community. Balbino Achabal ran the old Harrington Ranch and opened the Jack Creek Guest Ranch in 1943. Other Basque ranchers included Ysidro Urriola, Vincente Bilbao, Andres Inchausti, Joe Saval, and Feliz Plaza. Plaza also ran the Jack Creek store, which the Urriola family owned. Despite prohibition, the store also featured a bar serving bootleg whiskey and home-brewed beer. Travelers and sportsmen rented the upper floor of the two-story building. Urriola also had a small service station, a necessity in that remote area. Dances were a popular diversion at Jack Creek since trips to Elko were rare. Most people shopped by catalog. With the arrival of school buses, Jack Creek children rode to the Independence Valley school, and the Jackson District school closed.

Jack Creek continued to be a local sportsmen's mecca. The fishing there is some of the best in Elko County. Prominent Elko residents such as Harry Gallagher and Newt Crumley were frequent guests of the Urriolas. While owners changed, the so-called Jack Creek Resort remained open until the 1980s, providing lodging, food supplies, and gas for sportsmen and campers. Eunice and Loren Wilder bought the property in 1961, and while Loren died in 1976, Eunice remains a resident. Hank and Wendy Ispisua continued to run the business for their relatives until they closed it a few years ago. A number of ranches still operate in the area. The "resort" is still closed, but it could reopen in the future. Foundations mark the site of the power plant, and some parts of the flume are also still visible.

Lime Mountain

(Independence Mining District)

(Grand Junction Mining District)

DIRECTIONS: *From Jack Creek, head north on Nevada 226 for 8 miles to Lime Mountain.*

Prospectors from Cornucopia made initial discoveries on Lime Mountain in the late 1870s, but little production took place there, though the district was listed as a stop on the Northern Stage Company's daily stage from Tuscarora. In April 1871 John Ford discovered the Scottish Chief mine, which had initial assays of $307 a ton. He formed the Grand Junction Mining District, but the ore pocket quickly faded. The mine became active again during the summer of 1880, but no production was ever recorded. In 1875 there was a flurry of activity, and a number of mines opened, including the Syracuse, the Kentuck, the 74, the 75, the Britisher, and the Niagra. The Champion Mining Company worked the Champion mine. The Hero Mining Company owned the Hero and the Mountain View mines. However, interest faded when the Cornucopia boom ended, and in the 1880s Lime Mountain saw only limited activity. The Magnolia, the Daisy, the Mary Ann, the Scottish Chief, the Parrott, the Lime Mountain Consolidated, the Clyde, the Queen of the Hills, and the Deep Creek were all active at times but only produced a couple of thousand dollars worth of gold and silver. Jacob Eggers, sometimes referred to as Edgears, made the first significant discoveries in the area in March 1897, and the Eldorado and Queen of the Hills Consolidated were soon producing ore valued at seventy-five dollars per ton. The discovery generated little excitement, however, and few came to Lime Mountain to join Eggers, who was Elko County assessor and later served as Nevada State controller.

After the turn of the century, activity picked up throughout the district, and by 1910 almost forty miners were working around Lime Mountain. Eggers's company employed twenty-five people. Under superintendent C. J. Lyser, who built the first home on Lime Mountain, the company was working thirty claims. The Ibex Mining Company of Denver was also working claims, but activity faded by the early teens. By 1916 R. R. Young and Johnnie King had control of the Eldorado mine, which was renamed the Liberty, but little other activity was taking place.

Lime Mountain's most productive period began in 1928 when the Elko-Lime Mountain Gold Mines Company formed. The company made immediate plans to build two 100-ton mills, but the plans never materialized. Wood timbers for the mines came from old mines at Kennedy. By June 1929 eighteen tons of gold ore a day was shipped to the International Smelter in Utah. That year, production stood at 900 tons. Returns averaged $22.50 per ton, but the company folded in 1930. In 1931 Robert Clarke and R.T. Dougdale reopened the Liberty mine, now known as the Lime Mountain mine. The pair employed eight. Two six-ton trucks hauled ore to Elko for shipment to the American Smelter in Garfield, Utah.

Most of Lime Mountain's total production took place from 1947 to 1940. The Lime Mountain Consolidated Mining Company was organized in 1936 and began production from the Lime Mountain mine in April 1937. The biggest year was 1938. In that year, the area's mines yielded $106,700 worth of gold, silver, and copper, and the company generated revenue in the amount of $260,000. The company folded at the beginning of World War II, and since then there has been no production in the area. Since the war began, no exploration has taken place on Lime Mountain. Its total production stands at 8,000 ounces of gold, 24,000 ounces of silver, and 552,000 pounds of copper for an approximate value of $325,000. Scattered mining remnants mark the site of the Lime Mountain mines.

Lone Mountain
(Merrimac) (RipVan Winkle) (Phillipsburg)
(Florodora) (Nannies Peak)

DIRECTIONS: *From Dinner Station, head wêst for 4 miles. Take a left and continue 1 mile to Merrimac. To reach Rip Van Winkle, backtrack 1 mile and continue heading north for 4 miles. Head left for 1 mile and then south for 7 miles to Rip Van Winkle.*

First discoveries on Lone Mountain took place as early as 1866, and heavy prospecting occurred during the next few years, culminating in the organization of the Lone Mountain Mining District on June 18, 1869. The owners of the Gem Quality Lode met at the Tuolomne Saloon in Elko and named John Little as chairman, F.A. Rogers as secretary, and W.R. Litchfield as recorder. Reorganization of the district took place on May 25, 1870, with new additions to the bylaws. Robert Waddell was named new district recorder.

By 1871 a number of mines in the area were producing. They included the Whitlatch, the Lexington, the Freeman, the Mammoth, the Copper Queen, the Imperial, the Pauline, the Monitor, and the Humbug. Active mining companies included the San Francisco Mining Company, the B.N. Lowe and Company, the Wideman and Company, and the Earl, Rand and Company. The most promising early mine was the Pauline, which Lowe, McKenzie and Smith—who had also located the Monitor mine—had discovered. While a town did not develop, miners were scattered all over the mountain. In the early 1870s Henry Miller, who ran the Cornucopia stage road, built the Lone

Mountain Station, which miners used to ship ore and receive goods. New discoveries kept many prospectors combing the mounting during the next few years. New mines included the Whitlach Union, the Black Sail and the Red Lyon, the Constitution, the Caledonian, the Morningstar, the Evening Glory, and the Columbia.

Reorganization of the district took place on November 5, 1876, and it was renamed the Merrimac Mining District, after a prominent new mine. Production did occur on Lone Mountain in the 1870s but not at a sufficiently high level to make the district boom. With primary values in copper, lead, and silver, returns were fair but not spectacular. A new mine, the Buckeye, opened in 1879, and John Snymont, J. E. Galligan, and J. E. Wieland started the Carbonate mine. S. G. Weston, who would spend more than thirty years in the district, arrived on Lone Mountain in 1881 and became the biggest proponent of the Lone Mountain mines. In July 1881 he discovered the Comet mine, which he named after a comet that had appeared the day before.

Edward Reilly, owner of mining interests in the Railroad Mining District, purchased the Copper Queen mine in the summer of 1881. Other new mines opened, including the George Monitor, the Saint Helena, and the Mammoth. Trouble hit the district in August when water began entering the Merrimac mine, and forty barrels a day quickly began flooding the shaft. Over the next few years other Lone Mountain mines experienced similar problems, which caused production to slow down. A new mine, the Relief, became active in July 1883, and it helped restore flagging interest in the area. Other new mines discovered in 1883 included the Labyrinth, the Reno Consolidated, the Bona

Overview of the mill and camp at the Rip Van Winkle Mine, 1939. (Pauline Quinn collection, Northeastern Nevada Museum)

Fida, the Leopard, the Double M, and the Mollie. Joe Hall filed for a mill site, but the district's erratic production precluded the construction of a mill. In the next few years many of the old mines failed, only to be replaced by new ones such as the Micawber, the Staffordshire, the Kansas, the Bi-Metallic, the Cleopatra, the Iowa, the Langton, the Silver Dale, the Picollo, and the Silver Chief.

In 1887 Edward Reilly, George Bliss, and S. G. Weston discovered the King mine, and its ore ran thirty ounces of silver and five dollars of gold per ton. On January 1, 1888, Weston became district recorder. Prospectors continued to roam the canyons of Lone Mountain, establishing claims and digging shafts. By September 1890 Joe Mason had a seventy-five-foot shaft on his claim. Weston continued to explore and locate claims. He began working the Copper King mine, located at 10,000 feet, and a new vein of high-grade carbonate rewarded him in September 1891.

All this activity kept interest in the district alive, and 1897 proved to be its best year to date. The Merrimac, which Weston owned, was producing 18 to 20 percent copper with forty to sixty ounces of silver per ton. A big strike in the Copper Queen mine in June created new interest. Mason and Fleishman, who had prospected Lone Mountain for years, made a new copper strike. The first big mining sale in the district took place in March 1898 when Weston sold the Cuag Copper claim group to W. P. Bonbridge of San Francisco for $75,000. But Bonbridge defaulted on July 1 and Weston regained control. Shortly afterwards, Weston formed the Lone Mountain Mining Company to oversee all his mines. Winters on Lone Mountain could be harsh, especially at the higher elevations. In 1899 snow as deep as eighteen inches lay on the ground, and Weston had tunnels dug to connect the buildings and mines.

After the turn of the century, interest in Lone Mountain ventures continued to grow. Tasker Oddie, later Nevada governor, purchased an interest in a couple of mines, and in May 1900 the Baltimore group was bonded to A. H. Tarbet of Salt Lake City for $49,000. Tarbet organized the Pacific Consolidated Mining Company to work the claims. In May 1902 a strike made in the Morgan mine, which Weston owned, led to the immediate bonding of the mine and twenty-nine other claims to the Western Exploration Company.

During the spring of 1905 Charles Gilder and Charles Mayer's discovery of the Florodora (sometimes called Fledora) mine led to a small rush. In July Weston, ever the promoter, laid out the townsite of Florodora just below the mine and began selling lots, but the ore disappeared quickly and Florodora was just a memory by winter. In May of the following year G. M. Phillips, a veteran miner from Tonopah, discovered rich ore, and a new town named Phillipsburg sprang up. More than 150 lots were sold in the first week, and a saloon, a restaurant, and an assay office soon opened. Unfortunately, Phillipsburg's fate was the same as Florodora's. Phillips' mine played out, and everyone abandoned the town and mine by the fall. In July 1907 the Pacific Consolidated hit a good vein on the Baltimore group, but mines on Lone Mountain were more or less inactive until 1912. The value of Lone Mountain mines' production from 1870 to 1908 has been estimated at $80,000.

In February long-time local prospector John Yore discovered the Rip Van Winkle mine, which was destined to be the most productive one on Lone Mountain. The discovery of this mine, located on the west side of the mountain, revived interest in the district, and Yore immediately sold it for $25,000 and left Lone Mountain. He returned after squandering his money and located the Ajax mine, which proved worthless. In the summer of 1912 the Alaska Improvement Company bought the Rip Van Winkle mine for $38,000, the Copper Queen mine reopened, and the Pacific Consolidated expanded operations to the Baltimore group. The Ely Consolidated Mining Company purchased the Copper Queen mine in 1913 and was soon shipping 7.3 percent copper ore. In August 1915 Arthur Miller began shipping ore from the reopened Florodora mine and from a new mine called the Donna Roma, which had ore that assayed at $1,100 per ton in gold. This was the first gold found on the east slope.

Despite this activity, the district's main focus was the development of the Rip Van Winkle mine. The mine had new machinery and equipment installed in April 1916. The Elko Lead and Silver Mining Company leased the mine, which produced $45,000 worth of gold, silver, lead, and copper in 1916. A strong earthquake in 1917 shifted parts of the mountain, and most of the springs on the mountain slowed or shut off. In addition, extensive damage occurred in the deeper mines, particularly in the Rip Van Winkle. Repairs were not completed until 1918, when production resumed after the Alaska Development Company regained control of the mine. In 1921 Lee Hand discovered turquoise and developed the Hand mine into one of the leading turquoise mines in the United States. It held that rank until the veins ran out in the late 1920s. In the 1920s the Rip Van Winkle mine continued to produce most of the district's ore, although the Jack York, the Glider, the Jenkins, the Copper Queen, and the Pacific Consolidated mines also yielded small amounts.

John Yore, who had discovered the Rip Van Winkle, died on Lone Mountain in August 1930 while still searching for the Rip Van Winkle's companion mine that he claimed to have found but was never able to locate again. The early 1930s were quiet years for Lone Mountain, and the Alaska Company concentrated its efforts on further mine development. Most production during those years was a result of this development. A thirty-ton flotation plant at the mine, which was completed in November 1930, processed the small amount of treatable ore that was being removed. The Alaska Company sold all its holdings to the newly formed Rip Van Winkle Mining Company in September 1937.

A period of rapid expansion and large-scale production began on Lone Mountain, and the influx of people led to development in other areas. Earl Fordham, Walter Gann, Jack Myles, and E. A. Montrey began developing the Mesquite mine, and the Lone Mountain Mining Company was incorporated to work the Hunter claims. But it was the Rip Van Winkle property that really boomed. In January 1938 construction started on a 100-ton mill, a 22,000-gallon oil tank was bought from Robert "Doby Doc" Caudill, and a pipeline was built from Stewart Spring. In August the Utah Copper Company bought the property for $60,000 but retained the Rip Van Winkle Mining Company moniker. After the company had spent $100,000, the mill started up in September 1939 under the guidance of mill superintendent Power Warren. At the same time, the Rip Van Winkle Extension Company was organized to work the nearby Gilder mine, which H. W. Gilder had discovered in 1907. Two other mining companies that organized in September were the Malachite Gold, Silver, and Copper Company and the Sleepy Hollow Mining Company.

By January 1939 tent houses for a crew of thirty-two had been built, and by spring the number had risen to seventy. A school district was quickly established, and Merle Nutting taught in a tent house with a lumber floor. A frame school completed in September 1940 housed twenty students. By the end of 1940 there were eighteen cabins, fourteen apartments, and a cookhouse at the mine. The mine produced $282,300 worth of lead and zinc in 1940, and it ranked fifth in Nevada for production of these metals. On January 31, 1941, the Rip Van Winkle Mining Company and the Rip Van Winkle Extension Company combined to form the Rip Van Winkle Consolidated Mining Company.

In September 1941 Blawnie Mae Fairchild, who had taught throughout Elko County for a long time, began teaching at the Rip Van Winkle mine. She spent the week at the school and drove back to the Fairchild Ranch, near Tuscarora, for the weekends. The school included all grades, which made teaching difficult, but Fairchild was pleased with students' performance. One of her pupils, Jean Sabala, earned the highest S.A.T. score in the county in the year she took the test.

The Rip Van Winkle mine produced more than $152,000 worth of lead and zinc in 1941, and the mine was the second largest producer of these metals in Nevada, outranked only by Pioche. The following year was the mine's most productive, and it yielded $298,600 worth of lead and zinc. A drop to less than $144,000 in 1943 signaled that the Rip Van Winkle's productive years were soon to end. An additional blow compounded problems in May 1944, when three men were trapped in a cave in the mine. It was the first accident in the mine's history. All three men were extricated but one, Joe Modarelli, died a week later. During the summer of 1944 most residents living near the mine moved away, and by November the mill and mine were forced to close due to a lack of labor. Only $52,000 worth of lead, zinc, silver and copper was mined that year.

In May 1946 Preston and Rulon Neilson and Lowell Thompson reopened the property. They worked the mine until 1951. At the same time, the group was also working the Lone Wolf mine. The group's best two years were 1948, when it earned $34,000, and 1950, when it earned $23,000. After 1951 operations were curtailed, and only security guards were left at the mine. Only a small amount of production has taken place there since 1951, although the Lone Mountain area, which produced a total of 3.3 million pounds of zinc, still ranks as the largest producer of zinc in Elko County. Lone Mountain mines also produced 5.5 million pounds of lead, 728,640 ounces of silver, and 223,449 pounds of copper.

Much remains at Lone Mountain, although in 1994 a large fire destroyed many of the buildings left on the east slope. But there are still three groups of buildings at mines on the east slope. The remnants of the Rip Van Winkle mine, on the west slope, are the most substantial on Lone Mountain. Mill ruins dominate the site, and settling ponds and the few buildings remaining struggle to remain upright. The cookhouse has recently collapsed and lies across the main road. The school, which closed in 1944, was located on the flat above the mine, which is also where many miners' cabins and trailers were located in the 1930s and 1940s. Lone Mountain has a great deal to offer. With so many scattered mine sites to explore, plan a full day to enjoy it all.

Midas

(Gold Circle) (Summit) (Rosebud) (Dunscomb)

DIRECTIONS: *From Tuscarora head south and then west on old Nevada 18 for 41 miles to Midas.*

Midas and Jarbidge were eventually to become the biggest twentieth-century gold towns in Elko County. Midas was originally called Rosebud and then Gold Circle. James McDuffy discovered the first gold ore there in July 1907. He took his sample to a Tuscarora assayer, but though it proved valuable he did not file any claims. Shortly afterwards, McDuffy ran into an old friend, Richard Bamberger, who immediately came to the area and staked claims. McDuffy and Bamberger actively worked the Bamberger mine in September, but a disagreement ended the friendship soon afterwards, and McDuffy left. Initial assays from the mine ran between $2,400 and $20,000 per ton. This caused a rush to Gold Circle. In October Paul Ehlers discovered the Elko Prince mine, the district's best producer. The Elko–Tuscarora stage line immediately extended a branch to the camp, and within one month there were 150 people at Gold Circle.

Lots selling for $100 to $200 each in a townsite laid out in November went quickly. Gold Circle was the name selected for a post office, but it proved unacceptable because the word "gold" was part of so many other post offices' names. The alternative name of Midas, after the mythical Greek king, received a positive response, and a post office opened on November 16, with Patrick Leamy as postmaster. Other mines discovered by the end of the year included

Midas, 1908, looking south. (Enzo Gori collection, Northeastern Nevada Museum)

the Spotted Dog, the Esmerelda, and the St. Paul. The gold ore showed up in unusual places. C.D. Dorsey and Patrick Leamy, the postmaster, discovered a placer gold deposit while digging a well to supply the camp with water. It was the most valuable well ever dug in Elko County.

By the beginning of 1908 the Midas boom was on, and a small suburb, called Dunscomb after the local doctor, developed, but Midas soon absorbed it. By the end of January there were several saloons, two restaurants, a store, two feeding stables, and four real estate offices in Midas. Because of the high cost of shipping ore, only the highest grade ore was hauled to the faraway mills.

Another new discovery on Summit Creek, a short distance from Midas, led to the formation of another small boom camp. The Summit post office, with John Flavin as postmaster, opened at the camp on February 8. By March, Midas had a population of 350. The Gold Circle Townsite Company and the Summit Townsite Company became involved in a bitter fight for townsite supremacy. Two auto stage lines to Golconda were put into operation. N.F. Pressley ran the Pioneer Auto Line, which had four autos, while the Auto Transit Company ran two autos. Two stage lines also ran. On March 14 the Gold Circle Publishing Company began publishing the *Gold Circle Miner* in Midas. Mark Musgrove, who owned the company, had run many other newspapers throughout Nevada. The paper cost ten cents an issue. It did not last long and folded in May.

By the end of April, the population of Midas was estimated at 1,100. It continued to increase, peaking at about 1,500 by the summer. With the high value of gold ore, a keen interest in the town grew throughout the West. Newspapers made grandiose claims that Midas contained "the greatest gold zone ever discovered" and that the "gold is inexhaustible." The first big mining sales took place in the spring when J. A. Foley sold the Golden Crown mine to Willard Snyder, George Wingfield, C. C. Luce, and A. C. Ellis Jr. for $50,000. Soon afterwards, a group of New York investors bought the Elko Prince property for $35,000. Other active mines included the Midas, the Rex, and the St. Paul.

By the summer of 1908 many businesses had opened in Midas. Six saloons served the thirsty miners; they were the Northern, the Gold Dust, the Crystal, the Midas, the Capital, and the Eureka. Five hotels had been built, the largest of which was the Gold Circle Hotel, which could house fifty guests. Other hotels were the Hotel Nevada, the Grand, the Canyon, and the Monte Carlo. The Pioneer Restaurant and the Gold Circle Mercantile Company were also in business. The town even had its own slogan, Touch it and it turns to wealth. The Gold Circle News Publishing Company, which Frank Reber owned, began publishing a new newspaper, the *Gold Circle News,* on June 13. In July the Gold Circle Commercial Club was organized and soon had eighty

Midas in the 1930s. (Stanley Paher collection, Northeastern Nevada Museum)

members. Towards the end of the summer of 1908, however, Midas experienced a slowdown due mainly to the absence of mills in town. Only the richest ore could be shipped, and as a result many miners left town. The *Gold Circle News,* its revenues falling, folded in October. Despite drawbacks, however, mines around Midas earned their owners $37,000 in 1908, primarily in gold. But by the end of the year only 250 residents remained.

Midas had become prey to the boom and bust cycle that plagued many mining towns. While its mines produced every year from 1908 to 1943, the amount mined varied dramatically from year to year. The town stagnated in 1909, but production began increasing after the Rex Gold Mines Company, which Rothschild owned, installed a hoist at the Rex mine in April. It was the first hoist built in the district. The company was the district's biggest employer, with 100 on its payroll. Around this time a new mine, the Esmerelda, was bonded to H. H. Muggley for $100,000. Midas's first bank, the Morey, opened. The local mines' most pressing need continued to be mills, and in 1909 this need was finally addressed to some degree. In June the Rex Company completed a thirty-ton mill, and G. W. Hartley finished the twenty-five ton Tadmore mill, located at the mouth of the canyon. In addition, the American Ore Reduction Company completed a ten-stamp mill, and a four-ton mill started up at the Queen mine. By September the mines and mills employed 200 workers, and the Rex Company hired seventeen men to build a new fifty-ton mill. At the end of 1909 six mining companies controlled virtually all the district's production. They were the Gold Circle Crown Mining Company, the Midas Gold Mining Company, the Frying Pan Mining Company, the Rex Gold Mines Company, the Old Judge Mining Company, and the Eastern Star Mining Company. The district's total production for 1909 was valued at $90,000, and the largest producer was the Esmerelda mine.

In February 1910 the twenty-stamp Horton mill was completed, and it started up. Not to be outdone, the Rex Company completed its own new fifty-ton mill in March, but a tragedy marred its first run when Lee Gilson, the mill foreman, was killed while testing the new machinery. The Rex mine was now 200 feet deep, and it produced 1,000 tons of ore a month. A new mine, the Sleeping Beauty, began producing in 1910, and in May the forty-ton Lane mill was completed at the mine. Leslie Savage organized the Elko Prince Mining Company after purchasing the mine, which was to become one of the district's most prominent. In 1910 Midas's population held steady at around 200, and a new road was graded to Golconda.

The arrival of Haley's comet brought some excitement to the town, although a prank attracted everyone's attention away from the main event. A crowd gathered when someone saw a ball of fire on top of a hill. It began to descend to cries of "there's the comet." The "comet" turned out to be a bundle of underwear that a fun-loving resident had lit on fire and was carrying on top of a long pole.

Despite all its new mills, the value of Midas's production dropped to $25,389 in 1910. Over the next few years, the rate of production bounced up and down. In 1911 more than $60,000 worth of primarily gold was mined, but the value of production dropped to $40,000 in 1912 and to less than $35,000 in 1913. During this period the mills operated sporadically, and some closed down.

In May 1913 the Elko Prince property sold for $900,000. The Esmerelda was the area's largest producer in 1913. Two new mines, the Esperanza and the Hardscrabble, began production that year. In August 1914 a five-stamp, fifty-ton mill was completed at the Eastern Star mine, and the value of production rose to $68,000. F.J. Benneson took another stab at the newspaper business in Midas when he began publishing the *Gold Circle Porcupine* on May 13. There was little support for the paper, which was handwritten and mimeographed on old ledger paper, and it only lasted until May 30. In 1915 ten mines were being worked. The most productive of these were the Elko Prince, the Eastern Star, the Esmerelda, the Hardscrabble, and the Rex. After a couple of years of exploratory work, the owners of the Elko Prince property began to prepare for major production, and on December 13, 1915, they completed a fifty-ton mill at a cost of $80,300. Leslie Savage was the manager. A problem for both this and the Rex mill was the cost of ore treatment, which ran as high as eight dollars per ton. The price was so steep because of the quality of the ore.

Midas mines boomed again in 1916, and the period from 1916 to 1921 was the richest in the district's history. Almost half the district's total production took place in those years. The boom's main impetus was the completion of the Elko Prince mill. Three mills operated in 1916. A two-stamp mill was working at the Gold Circle Queen mine, and lessors reopened the Rex mill, which had closed in 1915. At the end of the year, cleanup at the Elko Prince mill yielded $70,000, and in 1916 total production for the district was valued at $347,107. Development of the Missing Link mine gave mining in the area a big boost.

By the beginning of 1917 only two mills, the Elko Prince and the Gold Circle Queen, were running, but the value of production rose to almost $400,000. A new mill, the Esmerelda, was completed on the Gold Circle Queen property, replacing the old two-stamp one. In November a new strike in the Rex mine, which Horton and Company of Battle Mountain now owned, occurred, and initial assays had values as high as $5,000 per ton. Production reached its highest point in 1918 when more than $456,000 worth of primarily gold was mined. Besides the strong values from the Elko Prince mine, the Big Chief Consolidated Mining Company, which owned the Missing Link and the Jackson Grant mines, enjoyed great success. Despite the fact that the number of active mines in Midas dropped from sixteen to six between 1917 and 1918, those remaining produced more than ever. But values began to drop in 1919 when the total value of the mines' production stood at only $343,000. This figure dropped to $207,061 in 1920 and to $204,400 in 1921.

On April 15, 1922, a gas tank at the Elko Prince mill exploded and fire destroyed the facility, but workers managed to rescue the large amount of gold contained in the precipitate tank. However, the company decided to curtail all operations because the value of storage ore piles and new ore was too low to make rebuilding profitable. Midas's glory years were over. In 1923 its mines produced only $2,200 worth of gold and silver, and the figures were even smaller for the years between 1924 and 1926. But while the production level was low, a great deal of activity was taking place on the investment front. In April 1923 the Red Hills Florence Mining Company, which George Wingfield owned, bought the General Jackson, the General Grant, the St. Paul, the Regal, the Gold Circle Run, and the Missing Link mines. In November the Noble Gold Mines, which Senator John Miller and R. T. Noble owned, took over the Colorado Grande claims. Noble Getchell, W. G. Adams, and H. V. Castle organized the Gold Circle Consolidated Mines Corporation in June 1924 and purchased the Rex mine and mill and the Weston, Casowick, and Camplain claims. Despite the prominent names associated with these new companies, the mines produced little, and the companies focused mainly on development work.

The Gold Circle Consolidated Mining Company continued to expand its holdings and in 1925 bought the Eastern Star and Elko Prince properties. A subsidiary, the Gold Domes Mines Company, formed to run the Elko Prince mine. Getchell's company then purchased the Missing Link, the Jackson, and the Grant mines in 1926. A setback occurred in August when a warehouse at the Elko Prince mine caught fire, destroying $8000 worth of new equipment, but the company continued to prepare for major operations. In March 1927 a rich strike in the Grant mine occurred, and a new 100-ton cyanide mill, designed by A. H. Jones of Salt Lake City, was completed in May at a cost of $175,000. Three shafts were employed to work the Elko Prince, the Rex, the Missing Link, the Jackson, the Grant, the Eastern Star, and the St. Paul mines. The mill produced its first gold brick on June 15, and Nevada governor Balzar attended the celebration. Around the same time, the Big Chief Consolidated Mining Company resumed work on a couple of mines but met with only limited success.

The best year of the new revival was 1928, when Midas mines produced $257,000 worth of primarily gold. The new mill was processing 2,200 tons of ore per month, but even though the value of production still stood at $164,000 in 1929, the revival had begun to fade. The mill closed down in August, and operations at most of the smaller mines were curtailed. Despite slipping production figures, the Gold Circle Consolidated Company continued to be active until 1940. During the 1930s a number of small companies and mills were active in and around Midas, and two mills were built in 1931.

The Young mill treated ore from the Esmerelda mine, and the experimental Hall Quicksilver Retort plant tried to recover mercury from the ore for awhile.

A celebrity amongst them created a sensation for the residents of Midas, when in September 1931 Jack Dempsey visited the town and returned a short time later to set up a training camp for an upcoming title fight. He began working a full shift in one of Getchell's mines, renamed the Champion mine. During his training period, Dempsey talked about becoming a partner in Getchell's ventures, and after the fight he returned and became part owner of the Champion mine, later renamed the Dempsey mine. Dempsey remained in Midas for a number of years and married comedy star Hannah Williams in Elko in July 1933.

Heavy snow in early 1932 caused many problems during the month of February. The old Boyle Grocery and Meat Market, considered a Midas landmark, collapsed, as did the Gold Circle Consolidated Warehouse. Another blow to the company came when the Missing Link mine's shaft house burned and some of the upper workings were destroyed. Despite these setbacks, the Midas chamber of commerce formed in August, and Noble Getchell was named president. In 1933 the Gold Circle Company reopened its mill and two new companies organized in June. The Midas Homestake Gold Mining Company controlled the Bonanza, the Reno, the January, the Betty Fraction, and the Superior claims, while James Foley, J.A. Barclay, and Jay Hansen formed the King Midas Gold Mining Company, which worked the Lucky Dog, the Red Dog, the New Deal, the White Dog, and the Jumbo claims. Because of all of this activity, Curtis True set up a motorized stage to Golconda and kept busy hauling businessmen and miners. In 1933 the Gold Circle Company was renamed the Gold and Silver Circle Mines and was listed on the New York Stock Exchange, probably because Getchell, now a U.S. Senator, was still the owner. A new seventy-five-ton concentration mill began operations at the Sleeping Beauty mine in December. East Standard Mining Company owned both the mine and the mill. The value of Midas mines production in 1933 stood at $86,785.

In 1934 the Midas District produced more than any other mining area in Elko County, but production value stood at little more than $50,000. The Elko Prince mine continued to be Midas's best and largest and now had a 900-foot shaft and 12,000 feet of drifts. The Buena Gold Mines took over the Esmerelda mine and rebuilt the old fifteen-ton Coots mill at the mine. In 1934 and 1935 the district earned more than $125,000, virtually all from mines the Gold and Silver Circle Company owned. By 1936 it was the only active company in the district. In 1937 the company treated 16,700 tons of ore, which generated $119,000. While the Elko Prince mine was still the leader, the Grant-Jackson produced almost as much.

While the ore produced was valued at $120,000 in 1938, and $139,000 above that amount in 1939, its value slipped to $72,000 in 1940. Labor shortages combined with decreasing ore values led the Gold and Silver Circle Company to shut down all operations in early 1941. This signaled the end of the mines' productive era. In July 1942 all the company's holdings were sold at a sheriff's sale for $10,000. The U.S. War Production Board prevented any resumption of mining operations at Midas, and the post office closed on September 30, 1942. It is interesting to note that after Isabel Lyon resigned as postmaster in 1929 there was no new master, and volunteers ran the office for thirteen years.

By 1950 there were only nine residents living at Midas, and the school finally closed in 1952. However, in the 1970s and 1980s Midas slowly revitalized, not because of mining but rather as a recreational community. Residents of Elko, Winnemucca, Reno, and Carson City, among other places, bought land and homes for summer retreats. A couple of saloons and a store cater to sportsmen, hunters, and permanent residents. The Midas Historical Society has been created to help preserve the heritage and original buildings left in the town. Some partial restoration has occurred at the long-neglected Midas Cemetery. Midas is a fascinating place to visit, and remnants of the prominent mining industry, which produced $4.1 million worth of gold and silver, are scattered over a wide area. In the town of Midas, many old buildings are intermixed with newer homes, and the school stands at the lower edge of town. There is much to see at Midas, but please respect the fact that the town is on private property.

Rio Tinto

DIRECTIONS: *Located 1½ miles west of Patsville.*

Rio Tinto was one of Nevada's last boomtowns. It existed solely because of the dogged determination of one man, Samuel Franklyn Hunt. In 1919 Hunt, who had prospected in places all over the west, including Alaska, found traces of copper a few miles south of Mountain City. He filed claims and spent the next twelve years telling anyone who would listen that there was a rich copper deposit under his claims. The name, Rio Tinto, which Hunt chose for his claims, illustrates his strong belief in his mine. The Rio Tinto copper mines in Spain are famous for having produced ore for 3,000 years. Hunt's main supporters during this time were the Davidson brothers, Walt and Jack, who ran a store in Mountain City. While most people laughed at Hunt, the Davidsons believed in him. They grubstaked him to whatever he needed, confident that their loyalty would be richly rewarded in the near future.

In 1930 Hunt's unflagging promotion of his mine finally paid off when

Rio Tinto Mill, 1930s. (Northeastern Nevada Museum)

One of the numerous apartment buildings constructed at Rio Tinto during the 1930s. (Northeastern Nevada Museum)

Ogden Chase became interested. Hunt and Chase formed the Rio Tinto Copper Company. With Hunt's glib tongue and Chase's mining promotion experience, the pair sold two million shares of stock, even though no mining had yet taken place. With the money they raised, they began work on an incline shaft. For years, Hunt had predicted that the copper ore body would be found at 250 feet. When he finally found copper on December 26, 1932, it

Overview of Rio Tinto during its peak years. (Stanley Paher collection, Northeastern Nevada Museum)

was at 227 feet. The ore assayed at an incredible 40 percent, which gradually rose as high as 47 percent.

Announcements of the strike started a rush to the area, which revived the nearby town of Mountain City and created two more towns, Rio Tinto and Patsville, the latter of which was located a mile below Rio Tinto. At Rio Tinto, a large group of tents and shacks quickly sprang up. The people who had originally bought stock at five cents a share saw their shares increase in value to $17.50 each. On June 30, 1932, the International Smelting and Refining Company, a subsidiary of the huge Anaconda Company, purchased the Rio Tinto Copper Company for $300,000 and renamed it the Mountain City Copper Company. To pay him back for all his support, Hunt named Jack Davidson the company's first president. During the next two years, the company focused its energies on developing both the mine and the town. It hired twenty-five men solely for the purpose of erecting buildings at the townsite. Wide, tree-lined streets were soon crowded with ten apartment buildings, duplexes, and many single-family homes. In addition, construction commenced on a grammar school and a high school, complete with athletic fields. A theater provided entertainment, and the town enjoyed modern conveniences such as electricity, running water, and a sewer system.

While this development was underway, a number of other companies filed claims in a ring around the Rio Tinto mine. In July 1933 the Warhorse Exploration Company bought thirty claims from Lyman Brooks and John Crosby and in August paid $500,000 for another block of claims. In March 1934 John Crosby, John Robbins, and Lois Hays formed four new mining companies. They were the Andover, the March, the Nipigon, and the Noji. However, the Rio Tinto claims covered the entire 200-foot thick and 700-foot long ore body, and there was little left over for other companies to find.

The Mountain City Copper Company began production in earnest in 1935 when it produced almost eight million pounds of copper. In 1936 a power line from Jarbidge provided power not only for the town but also for the new 300-ton mill that was under construction. A fatal accident marred Rio Tinto's rise to prominence, when deadly gas in the mine killed six men on August 13. Frank Teyera, Albert Apel, William Burns, June Barr, Lawrence Willis, and John Sheppard were all dead long before rescuers reached them. But the company town of Rio Tinto continued to grow. In September the mill, which had cost $300,000 to build, finally started up. This helped push Rio Tinto's copper production to more than 25 million pounds, worth $2.3 million, in 1936, making the company the largest copper producer in the United States. A post office, with Pearl Clary as postmaster, opened on December 7, and Rio Tinto's future looked bright. The next year was the town's best. Its mines produced more than 33 million pounds of copper, valued at $4 million. At

Two of the most complete buildings left in Rio Tinto. (Photo by Shawn Hall)

the time, there were 340 men on the company payroll, and the mine and mill operated around the clock.

On July 18, 1938, the *Rio Tinto News* made its first appearance. The debut issue included the following editorial:

> Humbly and yet with pride, we announce the birth of a newspaper, the *Rio Tinto News*. It has come into being because we know that a newspaper can be valuable to a community in a great many ways. It can help form and crystallize public opinion in progressive policies. It can make the citizens of Rio Tinto, Mountain City, and Patsville realize that these are now their home towns and that the business and social activities of their neighbors are a part of their lives and should be of interest to them. It can, and must, print reliable news and make itself a medium for doing good in the community. If it does all of these things sincerely, no matter how small its circulation, it will be successful in upholding the ideals of 'a good newspaper.'

The *Rio Tinto News* was a weekly. Harry Pazour, the editor, ran it off on a mimeograph machine. The paper, which cost five cents an issue, came out every Tuesday. There were only fourteen issues, and publication ended on April 21. The paper included no advertising, but it offered in-depth reports of happenings in the town, at the community center, and at the schools in both Rio Tinto and Mountain City. The final issue included the following closing statement:

With the copy of the issue which you hold in your hands, the *Rio Tinto News* brings to a close a life which, though short, has been a happy one. We hope that it has contributed something worthwhile to the community during its existence. That has been its purpose.

It is impossible to express in words the feelings that your support, your encouragement, your willingness to overlook shortcomings have occasioned in our hearts. It has made the work connected with putting out the paper a real pleasure. You are grand people and your kindness will never be forgotten.

The *Rio Tinto News* has been proud to be a part of the life of this community and it is its hope that there is much happiness in store for all of you.

In 1938 the mill was enlarged to 400 tons, but the value of copper produced dropped to $1.3 million. The Owyhee River Copper Company and the Rio Grande Mining Company began operations at Rio Tinto in 1938, but their success was limited. In February 1939 the Mountain City Mine and Mill Workers Union no. 466 was organized. It soon had 280 members who fought for a wage scale based on the current price of copper. The mill was enlarged once again, this time to 450 tons, and the value of the copper the mine produced rose to $3 million. While the onset of World War II hurt many mines in the West, the Rio Tinto continued to flourish because of copper's strategic nature. But while the mine produced $3.1 million worth of copper in 1940 and $2.6 million worth in 1941, the vast ore body was becoming depleted. Miners went on strike in October 1941 to protest against low wages and unsafe working conditions, but the strike ended the following month after Governor Carville made a personal appearance. The value of copper production dropped to $1.7 million in 1942, $1.4 million in 1943, and $1.3 million in 1944. The last year in which the mine produced copper worth over $1 million was 1944. During that year, the company laid workers off, and a sense of impending doom spread over the town. The mine produced only $650,000 worth of copper in 1945 and barely $100,000 worth in 1946. Actual mining ceased in the spring of 1947, and the company processed stockpiled ore, which yielded copper valued at $524,000. The remaining sixty workers were stunned when the company shut the mill down in September 1947. The town of Rio Tinto quickly emptied, and the post office, normally the last thing to die in a town, closed on February 29, 1948. Cleanup operations netted $90,000 over the next two years, but Anaconda pulled out its subsidiaries in 1949, and Rio Tinto became a ghost town. Most of the buildings were bought and moved to Elko, Carlin and Mountain City. A group of single-family houses from Rio Tinto still stands on Seventh Street in Elko.

In the late 1960s and early 1970s a leaching operation in Rio Tinto yielded another 7.7 million pounds of copper, but this did not revive the town. The

Loveland Construction Company dismantled the mill for scrap in 1979. Rio Tinto is a fascinating place to visit. A few small family homes and other buildings are still there. Foundations of the many houses and apartment buildings show the town's layout. The most impressive ruins in the town are those of the concrete school and the huge mill, just above the town. Rio Tinto is the essence of a ghost town and is a must-see for ghost town enthusiasts.

Swales Mountain

DIRECTIONS: *From Elko, head north on Nevada 225 for 12 miles. Exit left and head west for 9 miles to Swales Mountain.*

C.D. McNeill discovered copper and zinc on Swales Mountain in August 1912. He built a twenty-five-ton concentrator, but his attempts were not very successful and he gave up by 1915. New mines were not discovered on Swales Mountain until March 1927. John Fader and Pedro Obra drilled twenty-seven holes during their period of exploration, and John Yore, who had discovered the Rip Van Winkle mine at Lone Mountain, began working the Hilltop claims. He, too, achieved only limited success, however, and activity died within two years. Exploration took place during the next ten years, culminating when Earl Kain, Harry Collins, Nathan Bararoy, and Phil Gunshon bought the Swales Mountain Copper group in June 1937. They organized the Swales Mountain Copper Company, but their mines never produced anything. In May 1943 the Argelus Basic Copper Company reopened the old A.H. Berring lead and zinc claims. A.A. McClean was the company's president and J.P. McGlynn was supervisor of operations. The claims produced little. The Longshot mine, which produced some barite, and the Edgar Tourquoise mine, which produced some turquoise in 1993, represent the area's total production since then. Because of the limited scope of mining on Swales Mountain, little is left there. The most extensive workings are at the High Top claims.

Taylors
(Taylor Canyon)

DIRECTIONS: *Located 7 miles east of Tuscarora.*

Taylors, named in honor of Taylor Postlethwaite, was a stop on the Van Dreillen stage. A bed and two meals there cost $2.50. During the Cornucopia boom of the early 1870s, a new stage line joined the Tuscarora stage line at Taylors. The DeFrees mill, which was built in 1875 and which treated ore from the Tuscarora mines, was located at Taylors. W.J. Urton bought the mill in August 1880 for $971 at a sheriff's sale, but the mill saw little use. In 1898 the Dexter Mining Company of Tuscarora built a power plant that used water from Niagra Creek. The plant generated 10,000 volts sent via a wire strung to the Dexter mill. The company dismantled the old DeFrees mill and used the wood to build a boardinghouse and bunkhouse next to the power plant. A more efficient facility built later at Jack Creek, where the water supply was more reliable, eventually replaced the plant. A number of ranches are located at the mouth of the canyon. Ranch owners have included Ben and Nona Trembath, Rube Kilfoyle, Bing Crosby, and Willis Packer. Today, a bar and cabins cater to locals and sportsmen. The remains of the DeFrees mill are located to the east of the cabins.

Tuscarora

DIRECTIONS: *From Elko, head north on Nevada 225 for 27 miles. Head east on Nevada 226 for 17½ miles. Exit left on old Nevada 18 and continue 6.7 miles to Tuscarora.*

Though it was destined to become one of Nevada's legendary towns, Tuscarora's beginnings were humble, and it took almost ten years after the first discoveries were made in the area before it really began to expand. While many stories exist as to who made the initial placer discoveries in Tuscarora, it is generally accepted that a prospecting party set out from Austin and found gold placer deposits there in July 1867. Members of the party were probably Steve and John Beard, Hamilton McCann, Jacob Maderia, William Heath, A.M. Berry, John Hovenden, Charles Gardner, and Charles Benson, although there is some dispute as to whether or not John Beard was actually in the party. The party made its initial placer discovery along a creek named McCann Creek, after Hamilton McCann. On July 10, 1867, the party had a meeting at the Creek. Charles Benson suggested naming the new mining district Tuscarora, after the U.S. gunboat of the same name upon which he had served, and there was unanimous agreement.

The Beard brothers built an adobe fort at Tuscarora for protection against Indian attacks that never materialized. To provide water for continuous placer mining, construction began on two ditches, one to Gardiner Gulch from McCann Creek, and another from Three-Mile Creek, which was later extended to Six-Mile Creek. Despite all the efforts, no one became rich from placer mining, and only a small camp of prospectors formed around the Beards's fort and adobe home, called the Adobes. Efforts concentrated on placer mining. Only the Beard brothers investigated some lode deposits. In the summer of 1868 they brought a five-stamp mill to the Beard ledge, but it was not very successful.

In 1869 the lure of placer mining brought the first Chinese, many of whom

were now out of work after the completion of the Central Pacific Railroad in
the area. By the end of the year, more than 200 Chinese miners had arrived,
and they formed a Chinatown adjacent to the Tuscarora camp. The Chinese
were much more efficient at placer mining than the whites were. They purchased a large block of claims, organized into companies, and proceeded

with a high-volume placer operation. The twenty or thirty whites in the district greatly resented the success the Chinese enjoyed on the very claims they had sold, but the whites did not want to admit that they were not willing to work as hard as the Chinese miners did. In 1870 Tuscarora had a population of 119, of which 104 were Chinese and 15 were white.

Tuscarora was to undergo major changes in the early 1870s. The frustrated white miners branched out from the placer sites and began prospecting the surrounding hills. In 1871 W.O. Weed discovered silver ore on the slopes of Mount Blitzen, located about two miles north. He called his mine, situated at the future site of the new town of Tuscarora, the Young America. But the big boom at Cornucopia drained most of the area's miners, and little else happened during the next couple of years. At this time, most miners were not interested in silver but were rather looking for gold. Weed had to do placer mining to finance his exploration work at the Young America.

A post office opened at the old camp of Tuscarora on July 18, 1871, but closed again in October 1872 and reopened on April 16, 1873, with John Beard as postmaster. The first killing in Tuscarora took place on September 28, 1873, when Thomas Jones, a local farmer, fired at Hamilton McCann, a rancher who had been one of the discoverers of the Tuscarora placers. Jones had been gathering horses that were trespassing on McCann's property. When McCann tried to stop him, Jones shot him four times, and he died. There had been bad blood between the two for years.

Weed continued to work the Young America mine, and the thin vein of silver grew larger the farther the shaft went down. He bought the Beard brothers' old stamp mill and installed it at his mine. In July 1875 Colonel

Wooden headboard at the extensive Tuscarora cemetery. (Photo by Shawn Hall)

W.R. DeFrees, who was on his way to prospect at Cornucopia, stopped in Tuscarora and discovered a vein that was even richer than the Young America. He named his mine the Oroide and bought the Young America mill, which he moved to Taylor Canyon where there was a better water power source. In August Tom Hanoum discovered the Grand Deposit mine. In July 1876 J. Woods discovered a large three-foot vein of silver near the Young America mine. The find proved to be a true bonanza, and he named it the Grand Prize. The *Engineering and Mining Journal* reported that, "a rich strike is reported in this mine (the Grand Prize), which is situated on the Virginia ground, an old location made in May 1875. But we understand that no litigation is likely to grow out of that fact, owing to the existence of a spirit of compromise among the respective claimants. Some eighteen or twenty tons of rich ore from the new bonanza have been forwarded to the Leopard mill at Cornucopia for reduction, but the result of the workings has not yet been made public."[6] George Grayson immediately purchased the mine for $50,000 and organized the Grand Central Mining Company, which was later renamed the Grand Prize Mining Company. The first 100 tons of ore shipped to Cornucopia returned $33,400. While other mines and companies proved successful, the Grand Prize was always undisputedly the best in the district.

As soon as word spread throughout the West, a rush to the new camp of Tuscarora began. Old Tuscarora, which was located about two miles below the new camp, quickly emptied of whites. The post office relocated to the new town, but the large Chinese contingent remained at old Tuscarora, steadily working the placers. The new camp acquired some of the old town's buildings, and those left became Chinatown's center. Tuscarora's Chinatown, at one point, was second only to San Francisco's in size.

Even before the Grand Prize discovery, the new town's population stood at 150. The town contained twenty houses, four saloons, two hotels, two stores, and a lodging house. A tri-weekly stage ran from Tuscarora to Cornucopia. Following the Grand Prize discovery, Tuscarora boomed. The Grand Prize Company had purchased the twenty-stamp Windsor mill before it was finished, and the area received a big boost in February 1877 when it was completed. For the next year and a half, the mill produced an average of $2,500 worth of silver a day. By the summer of 1877 Tuscarora had a population of 3,000, which made it almost as large as Elko, the biggest town in the county. Although much of Tuscarora's population tended to be transient, there were 1,500 permanent residents. Businesses and new mining companies sprang up left and right. In May a separate town, called Linkton, began forming next to the Grand Prize mill. It was named for Samuel Linkton, the mill superintendent. Tuscarora's massive expansion during the summer gradually absorbed Linkton, whose townsite had only been platted in June.

Saloons were extremely popular in the mining town, and Tuscarora had plenty by the fall of 1877. They were the Progressive, the Silver Brick, the Pioneer, the C.O.D., the Young America, the Idaho, the Delta, the Empire, the Cabinet, the Palace, the Gem, and the Grand Prize. In addition, the Crystal Spring Brewery opened in the Bacon brothers' store. The town had several lodging houses and hotels, including the Star Lodging House, the Howerton Lodging House, the Wilkins Hotel, and the Gem Hotel. An abundant supply of attorneys served the town. Other businesses in Tuscarora included the Tuscarora Meat Market, the Pioneer Restaurant, the Virginia Chop House, the Tuscarora Blacksmith Shop, the Pioneer Boot and Shoe Maker, Dr. L. L. Davis, undertakers Darnes and Robbins, and auctioneers G. M. Roberts and Company.

Many stage lines sprang up in the area. Smith Van Dreillen operated a line from Elko to Tuscarora that continued on to Cornucopia, White Rock, Mountain City, and Island Mountain. While this was the main line to Tuscarora, demand was so great that the boomtown supported many other stage and freight lines. R. Barclay ran a fast freight line from Carlin, and Wardop and Barring ran the Tuscarora Fast Freight and Accommodation Line to Elko. Charles Haines ran a stage line to Battle Mountain, while Jack Gaston had one running to Rock Creek. Tuscarora also served as the terminus for the Northern Stage Company, run by C. L. Aguirre, that ran to White Rock and Mountain City. A jail, built in 1877, was not used much because serious criminals went to Elko's prison. For a town of its size, Tuscarora had a fairly lawful and peaceable citizenry.

Two newspapers began publication in Tuscarora in 1877. The *Tuscarora Times* started up on March 24 under the guidance of E. A. Littlefield, who also published the weekly *Elko Post*. The publication cost five dollars per year. On May 23 C. C. S. Wright began publishing the *Mining Review,* which initially came out twice a week but later became a daily. The two papers enjoyed such friendly competition that they merged on January 3, 1878, and the *Tuscarora Times-Review* was born. The daily paper's new owners were Oscar Fairchild and John Dennis.

In 1877 mining continued to reign supreme in Tuscarora. In June a new strike in the Grand Prize mine generated revenue in the amount of $152,000 in August alone. However, the Grand Prize experienced a setback in September when its hoisting works burned—a $30,000 loss. A new company, the Tuscarora Consolidated, organized, and T. S. Brown was named superintendent. The company owned the Warsaw, the Susan Jane, the Occidental, the May-be-so, and the Revenue mines. Other active mines in Tuscarora included the Young America (later renamed the Independence), the First West Extension, the Moscow, the Lida, the DeFrees (formerly the Oroide), the Navajo, and the Emmett. The Navajo sold in late 1877 for $17,000. In November the ten-stamp Independence mill started up and joined the twenty-stamp Grand Prize and the ten-stamp DeFrees mills. Because of the huge demand for fuel created by the mills, the lack of wood necessitated the use of sagebrush. Crews filled huge wagons with it and stockpiled it next to the mills. In 1877 the Tuscarora mines produced $900,714 worth of ore, bringing total revenue earned to $1.6 million. To the delight of the citizenry, the Tuscarora City Baths opened in December. Service clubs were also active in town, and the Masons, the Knights of Pythias, and the Odd Fellows all had lodges. Music was a popular form of entertainment. Groups such as the Tuscarora Brass Band, the Excelsior Brass Band, the Harmony Club, and the Young Ladies Orchestra emerged. A baseball team also formed. The boomtown, it seemed, had everything.

The two most productive years in Tuscarora's history were 1878 and 1879. In each of those years, Tuscarora's mines yielded more than $1 million worth of bullion. In January 1878 the district earned $206,752, making 1878 the best year ever, with production valued at $1.2 million. A new mine, the Belle Isle, discovered in 1879, began production. The town's permanent population settled at about 1,500, and because of the high volume of bullion being produced, Wells Fargo opened an office in Tuscarora in 1878. In 1879 the first water system was completed in Tuscarora. While it was woefully inadequate, it provided a wonderful service when it was running.

A bizarre incident, which took place in Tuscarora in 1879, attracted a great deal of interest. A local woman married a man known as Sam Pollard but left him shortly afterward, because she found out that he was actually female. Pollard's fellow miners had believed that he was a man. After the truth came out, Pollard made the most of the situation and gave lectures, dressed half as a man and half as a woman. Pollard, whose real name turned out to be Sarah, had devised the scheme when she was young to protect herself from her abusive father. Residents, while they were troubled by Sarah's disguise, generally accepted her.

By 1880 Tuscarora was well established and offered many amenities. Its school had 140 students, and Hattie Edwards and Alice Smith were the teachers. In January the Tuscarora Polytechnic Institute opened under the leadership of Professor J. F. Burner. L. I. Hogel, formerly of Cornucopia, organized the Tuscarora Jockey Club. Admission to the racetrack, located below town, was fifteen cents. Two new breweries kept the beer flowing in the town's many saloons. Otto Trilling built the Tuscarora Brewery in Navajo Gulch, and around the same time Frank Curieux started the Tuscarora Brewing Company. A great deal of confusion arose from the two establishments' similar names. A new mill, the Lancaster, started and began working on Argenta mine ore, and a new mine, the North Belle Isle, began producing. The value of production, while it was lower than it had been the previous two years, still stood at about $450,000.

Fires that had spared Tuscarora during the first few years of its existence began to plague the town. The preponderance of wooden buildings made this inevitable. On November 26 a fire completely destroyed T. M. Jones's Grand Prize Hotel, but he quickly rebuilt it, and the new hotel opened just before Christmas. While in 1880 the town's population, including the residents of Chinatown, still stood at 1,400, new discoveries in the Wood River region of Idaho started a small exodus from Tuscarora.

In 1881 the old Phariss building was reconstructed and converted into a new public school. By June 417 students in all grades were attending. New businesses in town included the Tuscarora Meat Market, the Tuscarora Drugstore, the Cornucopia Drugstore, the Pioneer Meat Market, the Tuscarora Beer Hall, the City Market, and Mero Fruits and General Produce. W. S. Hillman, the main undertaker in town, was also a furniture dealer and a house painter. Mine production slipped dramatically that year, when the area's mines produced only $151,000 worth of silver. This was a sign of things to come. Throughout Tuscarora's future, relative lulls followed highly productive periods. January 1882 was the first month since Tuscarora's mills had been built that no bullion was shipped. Despite this dramatic development, John Spence started a new Tuscarora–Elko fast freight line. A major fire hit the town in February, burning the express building, which housed the Wells Fargo, Western Union, and United States Stage Company offices. The fire spread and also destroyed the Tuscarora Meat Market building and William

Hoar's restaurant. This led to the formation of two volunteer fire departments, the Neptune and Confidence Fire Companies.

A new newspaper tried to challenge the powerful and revered *Tuscarora Times-Review,* which owner Oscar Fairchild had cut back to a weekly in 1881. The *Daily Mining News,* which Harry Fontecilla ran, made its first appearance in January 1883. Despite high hopes for its success, it never came out from under the shadow of the *Times-Review,* and it faded into literary oblivion on July 29. During the mid-1880s the big mines of the 1870s began to play out, and by 1883 Tuscarora's population had slipped to less than 1,000, though there were still more than 200 students in the school system. Mines yielded silver valued at $1.5 million from 1883 to 1887, but it was clear that Tuscarora had fallen into borasca. A small mill built at a new mine, the Dexter, in 1885, failed and was later dismantled in 1886. The Independence mine and mill closed in 1886 after producing silver valued at $606,000. In May the Grand Prize mine and mill closed. The town continued to suffer, and many businesses shut their doors. Wilson and McPheters' Elko–Tuscarora stage line was the only line left. The stages were full leaving town and empty on the return trip.

A shot in the arm arrived in October 1886 when the Nevada Queen mine, located in Squabble Gulch, was incorporated, with George Hickox as president and F. E. Coffin as superintendent. In addition, Steve Beard and A. Leichter organized the Tornado Consolidated Mining Company. The Grand Prize mine reopened, but the mill remained closed. In the town, the Tuscarora Drugstore, open since the early days, closed. John "Dutch" Brinkman bought the Tuscarora Brewery and came up with a new motto, Patronize home industry. A new post office was completed, and local grocer George Peltier was postmaster. But the town continued to shrink, and by the end of the year there were only 300 registered voters in Tuscarora.

In March 1887 the Nevada Queen was reincorporated as the Consolidated Mining Company. It was the district's best producer during the next four years, and it revived flagging interest in Tuscarora. Plans began to extend the Nevada Central Railroad from Battle Mountain through Tuscarora and to Idaho, but the plans never materialized. In 1887 businesses in Tuscarora included the Tuscarora Variety Store, the Capitol Saloon, the Gem Hotel and Restaurant, the Tuscarora Brewery, the Sideboard Saloon, the Cabinet Saloon, the A. B. Waller Drugstore, the Pioneer Meat Market, the Tuscarora Meat Market, the Palace Restaurant, the Gem Restaurant, the Nevada Saloon and Lodging House, the Bald Eagle Saloon, the Primeaux and Engler Store, the Chapman Lodging House, the Speth Jewelry Store, the George Peltier and Company store, the Occidental Lodging House, Chris Steffen "The Practical Bootmaker," and the Pacific Mutual Insurance Company. The town also supported a doctor, a veterinarian, and a photographer. In 1886 T. C. and Will Plunkett converted Plunkett's Hall, which was formerly a two-story lodging house with ten sleeping rooms, into the town's social center. It featured a roller skating rink, a dance hall, and a large performing stage.

During the summer of 1887 active mines in Tuscarora were the Belle Isle, the North Belle Isle, the Nevada Queen, the Navajo, the Commonwealth, the Grand Prize, the Diana, and the Found Treasure. For awhile R. M. Catlin, who would later build the Catlin Shale Plant in Elko, was superintendent of the Navajo mine. In June Mrs. Eliza Graham sold the Found Treasure for $40,000. A new company, the Pondere, was incorporated in August, and Charles was named president. Two other companies entered the district soon afterward; they were the Seal of Nevada Mining Company and the East Grand Prize Silver Mining Company. In October the Navajo-Independence and Grand Prize mills restarted after standing idle for a couple of years. The Grand Prize mill produced $50,000 worth of bullion in its first two weeks of operation. Three new mines—the Young America South, the Tuscarora Consolidated, and the Pondere—began production.

The mining revival had a positive effect on Tuscarora. Two new social organizations formed in 1888. In January the Mount Blitzen Boys Club began hosting dances and performing plays at Plunkett's Hall, and a few months later, Mrs. A. W. Brown organized the Women's Christian Temperance Union. That organization, needless to say, was *very* popular with the local saloons. In June Fred Bowman opened the first bottling plant, the Tuscarora Bottling Factory, that operated out of the Sagebrush Saloon, and Carrie Cady opened an ice cream parlor in the Masonic Hall. The prominent Tuscarora Guard organized, and a baseball team called the Tuscarora Clippers began playing. Players were W. B. Kink (catcher), F. Bierbricker (pitcher), Dan Williams (first base), J. E. Jack (second base), E. M. Chapin (third base), T. I. Irland (shortstop), Tom Francis (left field), E. Ross (center field), V. J. Tedley (right field), Will Plunkett (manager), and W. B. Kirk (captain). A former Tuscarora resident, "Wheezer" Dell, later became the first Nevadan to play professional baseball, when he played with the Dodgers for a couple of years in the teens.

In July 1888 the Tuscarora Water Company was incorporated and began to upgrade and replace the antiquated town water system. Tuscarora was blessed with plentiful water, but bringing it to the town efficiently was problematic. A completed twelve-mile iron pipeline brought water to Tuscarora, and by 1889 every house or business that could afford it had running water. In addition, fire hydrants throughout the town were a great help to the fire companies. In August Reverend E. F. Brown opened the Methodist, the first official church in Tuscarora, and in December the new Union (or Commonwealth) mill, a concentration facility, started up, but it proved unsuccessful and closed in August 1889.

In early 1889 the water company built an ice house, to residents' great satisfaction. W. S. Hillman, the undertaker, bought George Peltier's—one of the oldest stores in town. He renamed it the Tuscarora Trading Company.

In October the Elko and Tuscarora Trading companies merged to form the Elko-Tuscarora Mercantile Company and bought the old Woodruff, Ennor, and Williams store. Gardner and Son of Elko built a new school at a cost of $4,100. In November plans for the railroad began anew. They called for the Idaho, the Nevada, and the Colorado railroads to be based at Battle Mountain and to run to Tuscarora, Delamar, Silver City, and to the end of the line at Boise. The plans also included the expansion of the Nevada Central Railroad to standard gauge and the extension of the line to join the Carson and Colorado Railroad. The Nevada Central's superintendent, C. A. Hirchcliffe, made numerous trips to Tuscarora. The *Tuscarora Times-Review* flatly stated that it was a "done deal," but the grandiose plans soon fizzled away.[7]

In 1890 Tuscarora experienced both highs and lows. The once-great Grand Prize mine closed for good after producing about $2.6 million worth of primarily silver, all told. On the other hand, the district as a whole earned revenue of just under $700,000, which made 1890 the best year since the glory days of 1878 and 1879. A new mine, the Dexter, began production and would become the last prominent mine in the Tuscarora District. The stage line to Elko continued to operate and was reorganized as the Elko–Tuscarora Stage Company. The ice cream parlor, now run by Alice Decker and Annie Curieux, closed in September.

Tuscarora mines began to fade during the next couple of years, and many people left the area. The Young America mine closed in 1890, the Belle Isle in 1891, and the North Commonwealth in 1892. In March 1891 there was considerable excitement in Tuscarora when the old Young America shaft, which ran under Weed Street on the main road in Tuscarora, began to collapse, and gravel had to be hauled in to fill the large holes that developed. In July the first electric lights lit up some of the town's homes and businesses. By this time, the only mill still operating consistently was the DeFree's mill in Taylor Canyon. The Union and Grand Prize mills were leased and used only occasionally. All the mines were in serious financial trouble. Only the Coptis (formerly the Young America South) and the Independence had cash on hand. By September 1892 only the Commonwealth had a positive cash flow. The battle over the silver issue in Washington, D.C., did not help matters. In April 1892 the Tuscarora Silver Club formed to help promote silver as the base monetary unit, and 310 people soon joined. In March 1892 the Dexter Gold and Silver Mining Company organized, with W. J. Urton as president and superintendent. But fortunes continued to decline.

The years from 1894 to 1896 were the worst to date for production, and the impact weighed heavily on Tuscarora. In 1895 the Wells Fargo office, with little bullion to ship, closed down, and the Navajo mine, after producing $1.7 million worth of primarily silver, closed for good. In the same year, the *Tuscarora Times-Review* shrank to two pages and then folded on October 5. The Fairchild newspaper legacy was over. Oscar Fairchild, who had come to Tuscarora in 1877, died of heart failure in June 1897 at the age of sixty-seven. Both he and his son, Tracy, had run the *Tuscarora Times-Review* from its inception. Besides taking charge of the paper, Oscar had a dairy near town and served as postmaster for ten years. Before coming to Tuscarora, he had run papers in Placerville, Virginia City, Pioche, and Belmont, and had founded the *Reese River Reveille* in Austin. C. E. and E. L. Bingham revived the *Tuscarora Times-Review* when they began publishing the paper as a tri-weekly on June 15, 1897.

The paper's revival came about as a result of renewed mining activity beginning in early 1897. Production in the coming years never approached the volume seen during the 1870s boom and 1880s revival, but they were the last years during which Tuscarora's mines produced respectable amounts. The main impetus for this revival was the emergence of the Dexter mine. At this time, the Young America and the Nevada Queen mines reopened. In June the A. W. Sewell and Company grocery store opened. This was significant because the store eventually grew into a huge chain throughout northern Nevada. The Mayfair Market chain finally bought it out in the 1960s. Some of the town's old businesses had managed to survive through the lean years, and a few new ones opened. In 1897 Tuscarora's businesses included the Pioneer Meat Market, the Gear Saloon, Speth Jewelry, the Capitol Saloon, the Cottage Hotel, Fred Wilson Livery, the Antoine Primeaux Store (which had taken over the Elko-Tuscarora Mercantile Company), the Tuscarora Beer Hall, the Gem Restaurant, the Phil Snyder News Agency, the J. Hungerberg Barber Shop, W. S. Hillman (undertaker and furniture dealer), the Tuscarora Foundry, and the Chris Steffen Bootery. In December Joe, Tean, and Will Plunkett renovated Plunkett Hall. The floor space was now eighty by twenty-two feet, with a twenty-foot stage. The sixty-foot auditorium stood on three giant stringers that allowed the floor to be tilted as much as twenty-eight inches for theater performances and then returned to a flat position for dances and other functions.

In July 1897 the long-idle Navajo mill restarted and worked Nevada Queen ore. The Young America Gold Mining Company was incorporated in October, and W. B. Andrew was selected as president. A big boost came on February 1, 1898, when the new forty-stamp Dexter mill began operation. The mill joined the smaller Kinkead mill, which the company had built in 1894. The entire project cost $100,000. The power plant in Taylor Canyon could supply only enough power for twenty stamps. The company employed 125 men, and the mill was able to process sixty tons of ore a day. This boosted production and warranted the reopening of the Wells Fargo office in May. The Eira Mining Company completed a cyanide plant at the old Independence mine dumps in June and also reopened the Eira mine, which John Sutliff and George Smith discovered in the 1880s; but it produced little. The company, which was re-incorporated in February 1899 as the Tuscarora Chief Mining Company, also bought the old Navajo mill, refitted it, and started it in October. As a result of all this activity, the Tuscarora Miners Union was organized in October.

In January 1899 the second power plant for the Dexter mill started up at Jack Creek. For the first time, the mill was using all of its forty stamps, and the extra power the plant provided brought electricity to the town of Tuscarora. Heinrich, Brown, and Dern bought the Young America mine for $60,000 in February, and they purchased the old gallows frame at the North Belle Isle mine and installed it at the Young America. C. S. Ford began construction of a ten-stamp mill, which he completed in February 1900. The mill was one of the first facilities to run on gasoline. It cost only eight dollars to run the mill for twenty-four hours, but by March 1901 it was idle.

Tuscarora's population stood at 669 in 1900, but many people left during the next couple of years. The Plunkett family stayed, however, and Tom and Will took over publication of the *Tuscarora Times-Review* in November 1900. In November the Dexter and Tuscarora companies consolidated and formed the Dexter-Tuscarora Consolidated Mining Company, but despite all the outward signs of activity, the mines were beginning to falter again. Since the peak of the revival in 1898, when mines produced more than $185,000 worth of gold and silver, production had been steadily slipping. The mining companies explored feverishly, trying to find new deposits, but both the Commonwealth and the Nevada Queen mines closed for good in 1901. In October 1901 the old Grand Prize mine changed ownership. The new owners renamed it the Phoenix, but unlike the famous bird, it never came back to life. A further sign of decline was the scrapping and dismantling of the Grand Prize and Union mills in 1902. Due to the lack of good ore, the Dexter mine and mill closed in February 1903. It reopened in June, but about half of Tuscarora's population saw the writing on the wall and decided to leave nevertheless. During the revival from 1897 to 1902, area mines produced gold and silver valued at about $785,000. In this period, more gold than silver was produced, a dramatic departure from Tuscarora's early days. As a comparison, from 1902 to 1930, during which time mining had essentially ceased, the mines earned barely $447,000.

Even the venerable *Tuscarora Times-Review* was in danger. Tom Plunkett retired in July 1903, and his son, Will, tried to keep the paper going. He cut publication down to once a week, and the paper did not come out at all from October 20 to November 21. When it appeared again, it was with the following disclaimer:

> With this issue the *Times-Review*, which has been temporarily suspended for a while, enters upon a new career. Seeing no encouragement in the present, gloomy outlook of our camp to justify us in continuing the publication of a tri-weekly, we had about determined to abandon the field, but have been prevailed upon to continue, and henceforth (or until times

improve) this paper will be issued weekly. Saturday being our publication day. Owing to the fact that we have issued so irregularly since 15th of August, no charge will be made by us from that date to the present. The subscription price of this paper until further notice will be $4 per year in advance, six months $1.50, one month .50 cents. Thanking our patrons for their patronage during the past three years we hope to merit a continuance of the same. W. D. Plunkett, Editor and Publisher.[8]

The paper only lasted one month longer. It finally folded on December 26, and the last issue contained no mention of the fact that the paper's life had come to an end. W. D. Fairchild took the press and other equipment back and stored it at the family ranch for years. Robert "Doby Doc" Caudill later "acquired" it, and another valuable piece of Elko County's history disappeared forever. Will and Tean Plunkett purchased a store in Mountain City in 1904. They moved their house from Tuscarora and attached it to the store. The other Plunketts remained in Tuscarora for a few more years. Tom, an accomplished actor, spent his time performing Shakespeare to an empty house at Plunkett Hall, which was later scrapped for wood. Tean was the only one who remained in Tuscarora, running the Gem Restaurant. He died in 1943 when he attempted to rescue his dog from the burning building.

The Dexter mine's closure in 1903, after it had produced $1.2 million worth of gold and silver, marked the end of productive mining in Tuscarora. Water in the mine rose to within twenty inches of the top of the shaft. One of the main reasons the mine failed was that when the mill was built, the company manager and board of directors loaned money for the construction, and their exorbitant interest rate ate up all the profits, leaving stockholders high and dry. By the time the Dexter had closed for good, over fifty miles of shafts and tunnels crisscrossed underneath Tuscarora. Most of the town's remaining businesses closed or moved elsewhere. The brewery, which had been a mainstay in the town for twenty years, closed in 1905 because there was no demand for its product.

During the summer of 1906 eighty men worked to develop the Dexter and Commonwealth mines in the hope of finding more ore. The work culminated with the formation of the Tuscarora Nevada Mines Company in March 1907. The company controlled all the mines in Tuscarora. Two million shares of stock were issued, and the company announced plans to build a railroad to Elko. In July interest revived in the placers at old Tuscarora when the Nevada Hydraulic Mining and Milling Company hit a ten-inch vein in the old diggings. E. P. and P. C. Johnson, R. T. Noble, and W. D. Bray later released the claim, and a hydraulic plant was built. The town of Tuscarora struggled to survive. Two short-lived newspapers tried to make a go of it in 1907 and 1908. The *Tuscarora Mining News,* a weekly, began publication on August 17, 1907, but the paper was unsuccessful and folded in early 1908. *Business*

Talks appeared from September to November of 1908; however, it was not a regular paper but rather an advertisement sheet for the Mail Order Printing Company of Tuscarora and was published only occasionally.

A new mill, which John Pattison built, began working the dumps at the Navajo mine in July 1909. By 1911 the Tuscarora Nevada Company had enough ore to restart twenty stamps of the Dexter mill. At the same time, the company also built a 100-ton cyanide plant to rework old tailings. Despite these improvements, however, the company earned only $28,000 in 1911. In 1912 it ran up against financial problems, and work ceased. In April 1913 the company was finally forced into receivership and went bankrupt. After many delays, it sold all its property at a sheriff's auction in December 1915. One of the potential assets was missing because a fire that an arsonist had started burned the Dexter mill in December 1914. The loss was listed as $150,000. Given the circumstances, the timing seemed very odd. In August 1913 the Tuscarora–Elko stage line retired its last wooden coach. Passengers and freight still needing transportation now traveled by truck or auto.

Until the 1980s there were only sporadic attempts at mining around Tuscarora, and their success was seriously limited. Most of the activity was not new mining but rather reworking of old dumps. The Holden Mining and Milling Company built a fifty-ton mill in July 1919 and reworked tailings until the mill burned in 1920. In December 1927 Dan Zucconi completed a twenty-five-ton mill to rework tailings from the Independence mine. Estimates of the value of Tuscarora's total production range from $10 million to as high as $40 million. In 1918 the *Mining and Scientific Press* put the figure at $50 million and said that $10 million in dividends had been paid, but a more realistic estimate of the area's total production value appears to be around $15 to $20 million. It is likely that early production figures are higher than those that were reported, because the companies tended to manipulate numbers to avoid the bullion tax. About $1 million of the total came from placer deposits, most of which the Chinese mined from 1870 to 1900.

Over the years, many of Tuscarora's empty buildings were either moved or torn down. Since the 1920s the town's population has remained below fifty. A devastating fire that started in the home of Lovella Pendergast struck the town in July 1948. The fire destroyed the old express office and the Confidence fire station, along with seven other buildings along Weed Street, creating large gaps along the street. Only the timely arrival of the Spanish Ranch fire truck saved many of the other buildings.

In June 1966 Dennis Parks and his wife, Julie, came to Tuscarora and established the Tuscarora Pottery School. Students now come from all over the world to train under Parks. In the last fifteen years, mining has begun again at Tuscarora. A group called Tuscarora Associates reworked 400,000 tons of old tailings from 1979 to 1982. Microscopic gold provided moderate success for Horizon Gold beginning in 1987, and a large open pit mine began production just below Tuscarora at the Dexter mine site. Tuscarora received national attention when the open pit operation threatened the existence of the town. The residents protested and the expansion of the pit halted precisely at the town's border. However, the company could still proceed with the pit in the future, and the legal battle continues. Low gold prices and poor ore value forced Horizon to close down and remove its equipment. For the sake of Tuscarora, one can only hope that the closure is permanent. Unless the case is awarded to the residents, a rise in gold prices could see the town threatened again.

Today, Tuscarora is classified as a ghost town, even though there are still about twenty people living there. Some of the old buildings are occupied, and there are empty ones interspersed among newer buildings and trailers. For years, Della Phillips ran a museum in Tuscarora that offered an unparalleled collection of artifacts and photographs, but it is closed now. The post office is still in operation. There is also much to see on the outskirts of town. The ruins of the Independence mill are just west of Tuscarora. Many other mining ruins are scattered around the hills, although mining that took place later has destroyed quite a few of them. Just to the east of town is one of the most interesting cemeteries in the county. A large number of wooden markers remain, and some restoration is slowly being done. The cemetery contains quite a few unmarked graves, but many of the complete ones have beautiful and intricate headstones.

At the old Chinatown and the original Tuscarora townsite, little remains except some old cellars and easily visible placer workings. The last resident of Chinatown was Yan Tin, who died in 1927. In 1934 a group of boys from Tuscarora found $1,200 worth of gold dust and nuggets in the front yard of his cabin. At one time, as many as 500 Chinese were reported to be living in Tuscarora, but once the big boom hit in the mid-1870s they went from being a majority to a despised minority. In fact, an anti-Chinese society formed in the town. Many of the Chinese were forced into illegal enterprises to make a living, and opium dens were popular with both Chinese and whites. A couple of these dens, located in the sides of the hills, are still visible.

Tuscarora has always been the most recognizable mining town in Elko County, and it remains the queen of the county's ghost towns. Visitors to Tuscarora are guaranteed to enjoy themselves.

Jesse Winter Ranch at White Rock. (Roy and Ruth Roseberry collection, Northeastern Nevada Museum)

White Rock
(White Rock City) (Silver Creek)

DIRECTIONS: *Located 23 miles north of Jack Creek on Nevada 226.*

The Frost family first settled the White Rock area in 1870, and the Winter, Parus, and Riddle families followed shortly afterwards. Soon a number of ranches were scattered throughout the valley. The small town of White Rock quickly formed, more in response to nearby mining (for a complete history of mining near White Rock, please see the section on Edgemont) than to ranching, and Virgil Bartlett started the White Rock toll road in late 1870. Because of the mining taking place in the area, about twenty-five people lived in White Rock by 1871. A post office, with Roger Robinett as postmaster, opened on June 21, but closed on October 3, 1872. It reopened on May 9, 1873, and remained in operation until 1925. The Tuscarora–Mountain City daily stage came through White Rock. When interest in mining near White Rock was at its peak, two other stages came to the town as well. One belonged to Haes and Dolin and the other to Van Dreillen. C. S. Anderson, who later ran Taylor Station near Tuscarora, opened a small mercantile store that served as an outlet for local farmers and featured such specials as a dozen eggs for twenty-five cents. A sawmill satisfied the demand for lumber. It provided timbers for both White Rock and Cornucopia. A school opened in the mid-1870s.

When interest in the nearby mines faded in the late 1870s, White Rock became entirely dependent on local ranchers and farmers. Fears of Indian attacks in 1876 forced White Rock residents to Cornucopia until the tension dissipated. In 1877 a constable, L. Orwfiler, and a justice of the peace, V. J. Bodrette, began serving the local populace. Bodrette became well-known in the county when he invented a cricket-killing machine.

While there were many ranches in the White Rock area, the most prominent was the YP or 25 Ranch, which the Garat family ran. John Baptiste and Gracianna Garat originated the YP brand in 1852 in California. It was one of few patented brands and is believed to be the third oldest in the country. The Garats were French Basques who came to White Rock in 1874, when they purchased 320 acres from Captain Stiles. It was the beginning of a four-generation ranching tradition and one of the largest ranching empires in Elko County. In 1895 a son, John, took over the ranch and married Matilda Indart. Peter Indart (Matilda's brother) and John Arrambede became Garat's partners. The dreadful winter of 1889–1890 was particularly devastating for the YP Ranch. The owners lost more than 10,000 head of cattle, but immediately began working to replenish the herd. In the 1890s the owners of the YP built a huge home, which later moved to the SL Ranch when it changed hands in 1910. John Garat's sons, George Henry, John III, and Charles, gradually began to run the ranch for their father. At its peak, the Garat family owned the original YP, the SL, the T-J, and the Meecham (or Mitchum) ranches, in addition to a few smaller ones. The family sold its ranching empire in March 1939, shortly after John Garat II died. Petan Land and Livestock Company, which Pete and Ann Jackson owned, bought it. The purchase price of $850,000 included 75,000 acres of land, 8,000 head of cattle, and 500 horses.

During the 1980s the Jesse Winter Ranch served as the ranching community's social center at White Rock. Winter had built a huge stone and log home with eight large, comfortable rooms. The school, the store, and the post office were also located at the ranch, which was essentially the "town" of White Rock. While White Rock's population stood at only thirty-three in 1880, the 1880s saw a substantial increase in the number of ranches in the area. White Rock became the terminus of the White Rock branch of the Northern Stage Company's daily stage from Tuscarora. The fare to White Rock was five dollars. The town also served as the terminus for the Owyhee Stage Company, which local resident Homer Andrae ran. W. D. Fuller brought the mail to White Rock from Tuscarora twice a week. By 1886 there were sixty registered voters at White Rock, and John Frost served as the precinct's voter registry agent. Frost was also the local road superintendent. Harry Riddle was named the new constable, and A. J. Sinclair was the justice of the peace. The school continued to operate at the Winter Ranch. In 1889 teacher Lottie Chaplin had six students: John, Laura, Willie, and Elvis Winter; and Florence Garst and Victor Parus.

In 1891 mail service for Owyhee to White Rock ceased and the route was abolished, leaving White Rock dependent on Tuscarora for mail delivery.

As became apparent in July 1892, the stage's reliability was questionable at best. During that month, a stage left Tuscarora carrying mail and a shipment of alcohol consigned to a local rancher for lubricating purposes during the haying season. When the stage did not show up at White Rock, a search began. The search party found the team unhitched and wandering around, and while no harm had come to the mail, the alcohol was gone. The search for the missing driver, Mort Groton, was fruitless, but a couple of days later he wandered into Columbia in a state of complete drunkenness. Because of his behavior, the company fired him.

In the mid-1880s John Frost bought the Storff Ranch and opened a store, which W. H. Wibra ran. Frost's wife died in June 1898 after falling from a ladder. Frost, the first settler in White Rock, left the area soon afterwards. By 1900 White Rock's population was 132. The town supported a store and a saloon and had a completed telegraph line in 1903. A fire at the Jesse Winter Ranch in April 1916 destroyed all Winter's barns, sheds, outbuildings, and machinery, as well as forty tons of hay. True to the spirit of the West, local residents held an old-fashioned barn raising and rebuilt the burnt structures.

A murder shocked the community of White Rock in October 1917. Arne Parus and John Winter had been embroiled in a long-standing feud that had started when Winter beat up Parus's father, leaving him barely alive, years before. Because the two men's brands were very similar, they continually charged each other with brand altering. Not long before the murder took place, Winter attacked Parus with a pair of brass knuckles. The feud came to a head when, during a roundup, a calf with Parus's brand followed a cow with Winter's brand. Parus shot the unarmed Winter five times, and Winter was dead before he could be brought into the house. The case did not make it to court until 1919 because of the influenza epidemic that plagued many areas of the United States in 1918. In May the jury stood eleven to one for conviction, and a hung jury was declared. The lone objector was J. V. Lockhart, who believed Winter had been going for his gun when Parus shot him. A new trial in October ended quickly when Parus was found guilty of manslaughter and sentenced to two to ten years. Parus, who became a trusty at the state prison, drowned on June 1, 1923, while herding cattle across the flood-swollen Carson River. Soon after Winter's murder, his wife sold the ranch to John Reed. Parus's brother, Raymond, continued running the ranch until he died suddenly while working his fields in 1948. The Petan Company bought the Parus Ranch in 1950.

In the 1920s the Garat family bought a number of smaller ranches. About three years after Jesse Winter died in 1922, his son, Henry, sold his father's ranch to the Garats. This purchase made the Garats's the largest cattle company in Elko County. Because of declining revenues, the White Rock post office closed on December 31, 1925. Shortly afterwards the store and saloon also closed. The school at the old Winter Ranch shut its doors in the 1920s.

Most of the ranches currently in operation around White Rock are part of the huge Petan Company. At the actual White Rock site, which is located on a private ranch, only foundations remain. Nearby ranches still use some of the old ranch houses. The original Garat home is now the Petan Company headquarters.

North Central Elko County

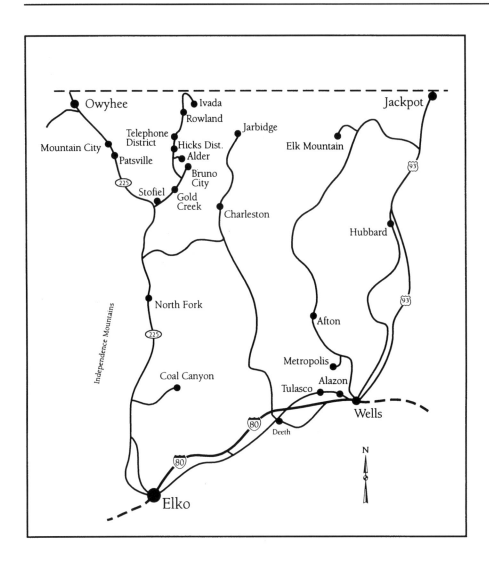

Afton
(Taber City)

DIRECTIONS: *From Metropolis, head north for 11 miles to Afton.*

Afton was born during the large Mormon dry farming experiment. Its first settlers—most of whom were from Afton, Wyoming, where another dry farming attempt had turned to dust during an extended drought—arrived in 1910 and 1911. By 1914 fifty homesteaders had begun to work the land. Afton's lifeline was Metropolis, the hub of the dry farming district and the source of all farming supplies.

A post office opened on September 26, 1914, with John Luckart as postmaster. His wife, Ada, took over the job in 1915 until the office closed on January 15, 1918. A fire destroyed the Luckarts' home and with it the post office in September 1916. They built a new home that also housed the post office.

A small extended settlement developed along Taber and Pole Creeks, with Afton as its center. The area was also known as Taber City. It was named after Henry Taber, who homesteaded there in the 1870s but moved to Elko in 1883 when he was elected Elko County sheriff. By the mid-teens, prominent Afton families included Ann and Inga Engh, Frank Truitt, and Frank and Tillie Powers. The community was tightknit and religious. The Taber City school opened in 1915. Mrs. Moses was the teacher. William Hutchinson later took over and taught for many years. Although many people had left Afton by 1920, the school remained open until 1933.

For a couple of years after the first settlers arrived in Afton, enough rain fell to produce substantial crops. By 1914 farmers had planted close to 600 acres with dry land wheat, which brought more than twenty bushels to an acre—a much higher yield than dry farming areas at Metropolis, Tobar, and Decoy could claim.

The year 1915 was extremely dry, and at Afton the lack of rainfall was compounded by an invasion of rabbits. Crops failed completely at twelve farms, while another ten suffered severe damage. In 1916 hard June frosts exacerbated the problems. By 1917 only half the original number of settlers remained at Afton. While conditions eased in subsequent years, just a handful of families were still farming in Afton by the 1920s. The area is still used for farming and raising livestock, and descendants of the original settlers own most of the existing ranches. To this day, the Enghs hold family reunions at their old homestead.

The remains of numerous old family homesteads dot the area around Afton. The site of Afton, where the school and post office once stood, is marked by foundations, scattered wood, and a new well.

Alazon

DIRECTIONS: Located 3½ miles west of Wells.

In 1909 the Southern Pacific Railroad straightened its tracks between Deeth and Wells, and Alazon came into being. The town marked the junction of the Western and Southern Pacific Railroads with a section house and worker homes. As many as twenty people lived at Alazon at times, and a small school, which remained in operation until 1938 when the district consolidated with Wells, opened during the mid-teens. Alazon also served as a telegraph station and was the first stop west of Wells. During the spring of 1942 floods plagued Elko County, and Alazon was completely isolated for several weeks after the only road washed out.

A murder took place in the little town in October 1948, when Richard Boudreau shot his friend, Richard Stewart, an Alazon section worker at Mineral Springs, north of Wells. John Moschetti, manager of the Thousand Springs Trading Post at Wilkins, discovered his abandoned, blood-spattered car with the body a short distance away. Circumstances surrounding the shooting remain murky, and no motive was ever established. Boudreau, also known as Dick Days, was sentenced to die on December 17, 1948. His sentence was later commuted to life imprisonment when Stewart's mother appeared to plead for mercy on Boudreau's behalf.

Alazon continued to serve the Southern Pacific until August 1956, when the railroad moved its offices to Wells. At that time, three full and relief operators and their families lived in company homes at Alazon. This was the end of Alazon. The railroad removed all remaining structures during track reconstruction. Nothing is left of Alazon today.

Alder
(Delores) (Tennessee Gulch)

DIRECTIONS: From Gold Creek, head north for 7 miles. Exit east and follow the road for 1½ miles to Alder.

Discoverers first came across placer in the Alder District on Young American Creek in 1869. They named Alder after a mining district they had worked in Montana in the mid-1860s. They organized the Alder Mining District, and limited placer mining took place there during the next few years. However, after no substantial deposits were found by the mid-1870s, the town was abandoned.

During the summer of 1885 interest in Alder revived when Henry Freudenthal, later sheriff of Lincoln County, discovered the Gold Bug mine, which he sold to James Penrod. Around the same time, Louis Burkhardt discovered the

Golden Gate mine. In November 1891 Emmanuel Penrod, who was instrumental in Gold Creek's rise to prominence, discovered the Oro Grande—the most productive of Alder's mines. It produced small amounts of placer, but it was not until 1895 that Joseph Lang, M.J. Penrod, and James Penrod's discoveries renewed curiosity about Alder's potential. The Alder Mining District was reorganized on August 29, 1895, with Emmanuel Penrod as chairman and Louis Burkhardt as secretary and recorder. By 1896 Tennessee Gulch consisted of four active mines, including the Oro Grande (also known as the Browning Brothers) and the Last Chance (also known as the Jack Abel).

In January 1897 seventeen new claims became active, and close to thirty people set up residence in crude homes at Alder. Prospectors began work on another twenty-five claims in February, and a townsite named Delores was platted. Residents submitted to the government a thirty-one-signature petition for the establishment of a post office.

The Oro Grande mine owners turned down an offer of $20,000 for their mine in 1897, but Jack Abel and Louis Burkhardt bought the Sunset claim group for $50,000. The Gluck Auf, or McDonald, began production in 1898, and with all its active mines, Alder gave the appearance of a bustling and permanent mining camp. Emmanuel Penrod built a three-stamp mill at a cost of $5,000. It began production on May 1, 1898, working Gold Bug and Oro Grande ore. However, excitement soon faded when prospective investors learned that initial assay reports were misleading, and ore values dropped quickly. The mill remained in operation for only a few months before being shut down. By the turn of the century, no one was left to object when the government finally rejected the Delores post office application.

Limited mining did take place at Alder from 1916 to 1939, when the Clipper Alder, the Pittsburg, and the Mohawk mines were active and the refurbished Penrod mill saw occasional use. But Alder's twenty-three years of mining activity produced only 53 ounces of gold and 1,219 ounces of silver. In 1949 the newly discovered Garnet mine produced modest amounts of tungsten. Union Carbide drilled there extensively and defined a large ore body. It shipped ore to a mill in Mountain City. From 1970 to 1977 the mine produced 6,000 units of tungsten.

Little activity has occurred at Alder since 1977, and only sunken cellars and ruins of Penrod's mill remain to mark the Delores townsite. The scenery, however, is spectacular, and a campground is located nearby.

*Head gates for the large
water ditch along
Crystal, Martin, Creek.
(Photo by Shawn Hall)*

Bruno City
(Bruneau City) (Wyoming District)

DIRECTIONS: *From Gold Creek, head north for 2½ miles. Turn right
and follow the road for 1 mile to Bruno City.*

Bruno City formed in July 1869 when the Young America mine
was discovered. The town was located three miles north of Gold Creek on
present-day Martin Creek. This site should not be confused with the Bruneau
mine, located farther north, near Rowland. Initial discoveries in the canyon
were made in 1864, but the White Pine rush to Hamilton drained manpower
and kept any development from occurring. The mining district was not organized until November 3, 1869. An adjacent mining district, the Wyoming,
was also established but was later absorbed into the Bruno district.

Early French settlers named Bruneau Canyon, but denizens of Bruno City
corrupted the name. Grazing had been taking place in Martin Canyon since
1867, when Hugh Martin and his family moved from Mountain City to the
canyon named after him. Pete Columbet and Dan Murphy also used the area
for cattle and horses. Initial returns from the Young America ranged from
$300 to $600 per ton. In November the Bruno Ledge started developing and
returned $384 per ton. Jerome Cross, one of the discoverers and a former
conductor for the Central Pacific Railroad, left and made his mark mining
near Salt Lake City.

By 1870 the census recorded 122 residents in Bruno City, and businesses
including a couple of saloons, stores, and a two-story hotel opened. Many

mines in the district were active by the end of 1870. George Washington Mardis, who later discovered gold at Charleston and Island Mountain, owned the Mardis and Miner's Rest mines. Mardis, Will Rogers, John Brooks, and A.D. Meacham found the Mardis mine in June 1871. Savage, McMillan and Company ran the Miner's Delight, the Madison, and the Sonoma, while Klinck and Company owned the Mountain King and Doane and Company ran the Big Giant. Other active mines included the Rosebud, the Saratoga, the Senator, the Humbug, the Fairview, and the Chrysopolis. In June 1870 J.A. Norton and Timothy O'Flagherty discovered the Argument mine, named after O'Flagherty shot Norton in the leg during a dispute. Emmanuel Penrod, who had eighty acres of placer ground, built a water ditch from Martin Creek to the head of the east fork (known as Fuzzi Gulch) of Gold Creek. Others dug a number of additional water ditches in order to work their claims. Unfortunately, the water only flowed at full capacity during the spring. Penrod ended up working the claims until 1892, while also working placer claims in the nearby Island Mountain district.

The Mardis Silver Mining Company formed in Chicago in January 1875 and purchased the Mardis mine. Although there were many mines in the vicinity, production was slow because of the remoteness of the area and the resulting difficulty in shipping ore. In July 1875 that problem was solved when the Mardis Company bought the ten-stamp Golden Age mill at Silver City and moved it to Bruno City. It took almost a year to move and install the mill, which started up in June 1876. By then ore values had fallen and the mill had little ore to process. The facility was sold to Henry and Robert Catlin, builders of the mill, at a sheriff's auction on August 24, 1877, and they sold it to Hiram Hyde in July 1878. People left, the mill and mines were idle, most businesses closed, and by 1880 only twenty people remained in Bruno City.

The final blow came when George Mardis was killed in a robbery near Gold Creek in September 1880. On November 3 the local miners held a meeting and selected Hugh Martin as the new district recorder, replacing Mardis. At the same time, they changed Crystal Creek's name to Martin Creek. Soon afterwards, the Mardis Company, Bruno City's main source of life, folded.

Some minor activity took place in the area during the next fifteen years, but no one ever resided at Bruno City again. In October 1884 a seemingly promising new discovery was made in the Lady Anna mine, formerly the Young America, which C.M. Moore, A. Leichter, and T.H., C.A., and G.S. Watkins owned. However, the vein pinched out quickly. In July 1893 J.A. McBride, who later became the first mayor of Elko, and Emmanuel Penrod found a new, rich placer deposit. The sand yielded a couple of hundred ounces of gold before the deposit was mined out in 1895. During the revival in the Island Mountain district in 1897 the Catlin brothers reopened the mill at the junction of Martin and Mill Creeks, but after experiencing limited success, closed it down for good a year later. In 1902 the Deer Creek mine became active. Under the management of Art Carlson, a seventy-five-ton mill began producing in January 1903. However, optimism about the mill's success was unwarranted, and the mill and mine closed in September.

While figures are sketchy for the early years, it appears that the district earned around $100,000 from the time of its discovery until the 1950s. From 1954 to 1982, the Diamond Jim mine, located on the old Mardis claim, was responsible for 95 percent of the district's total production. More than 1.6 million pounds of lead, 120,000 ounces of silver, and 83,000 pounds of zinc were mined.

Not much remains of Bruno City. The mill foundations, located just above the Gold Creek Ranger Station, are the only visible signs that the town was once inhabited. Some sunken cellars and scant, scattered wood mark the townsite, and the huge head gates for the main water ditch are located further up the canyon.

Charleston
(Mardis) (Bayard) (Union Gulch)
(Cornwall) (Copper Mountain)

DIRECTIONS: *From Wild Horse, head south on Nevada 225 for 8 miles. Exit left and follow the road for 25 miles to Charleston.*

Mardis, later renamed Charleston, came into being after George Washington Mardis discovered placer gold in 1876 on aptly named 76 Creek. Mardis had been active in the area since the late 1860s, and his company, the Mardis Silver Mining Company, was prominent in the Wyoming Mining District, which was later renamed the Island Mountain Mining District. While discoveries on 76 Creek led to the creation of the town of Mardis, mines in the district, including the San Francisco, the Black Warrior, the Sherman, the North Star, the Eclipse, the General Grant, the Commodore, the Chicago, and the Golden Gate had been active since the early 1870s.

The town of Mardis grew quickly and soon contained a number of stores and saloons, a hotel, an icehouse, a school, and a widespread reputation for lawlessness. Mardis, William McLaughlin, and Johnny Brooks had actually organized The Mardis Mining District in the spring of 1872, but the active mines were located along the Bruneau River, not at the future Mardis townsite.

Mardis, also known as "Old Allegheny," is one of the more intriguing characters in Elko County history. He was raised in Pennsylvania and talked constantly about the Allegheny Mountains there. His appearance was intimidating, because an accidental mining explosion had taken one of his eyes and

Ore house and bin at the Mission Cross Mine, located north of Charleston. (Photo by Shawn Hall)

scarred and blackened one side of his face. But Mardis had a reputation for honesty and fairness, and everyone who knew him trusted him.

Mardis became a bible-quoting preacher following his involvement in an incident at Mountain City. While drunk, he got in a fight. He knocked his opponent unconscious, and the crowd convinced him that he had killed the other man. Panicked, Mardis disappeared for a couple of weeks. After he learned of the ruse he was never known to drink again.

Mardis hauled ore from many northern Elko County mines to Deeth and Elko, and miners trusted him to carry their gold to banks in Elko. In September 1880 he made a trip to Elko carrying miners' gold to deposit, as well as $250 to buy supplies for Gold Creek's Chinatown. After traveling for some time, a Chinese man known as "New York Charley" sprang from the brush, startling Mardis and his team, and demanded Mardis' money. When Mardis told him that all he had was some chewing tobacco, Charley shot him four times. Mardis was not dead, however, so Charley finished him off with his knife. Mardis' body was found a few hours later, and a posse formed at Gold Creek. Charley had inexplicably taken off his shoes when he fled, and his six-toed footprint left a plain trail for the posse to follow. They found him hiding in Gold Creek. Because he had stolen his own people's money, the Chinese community pleaded to mete out their own brand of justice to Charley. The posse granted their request, and they dealt with him quickly. His funeral took place at the same time Mardis' did, and both were buried in the Gold Creek Cemetery.

The town of Mardis faded quickly after Mardis' death, and by 1883 only a handful of people remained there. Mines (the Troy, the War Eagle, the Smokey City, the Sunnyside, and the Palisade) were still active during the early 1880s. Placer deposits were also active, but their success was limited.

In 1885 the Mardis Mining District reorganized, and Isaac Sherwood was named recorder. Sherwood served as justice of the peace and was also postmaster for the Bayard post office, which opened at Mardis on August 28, 1886. The residents appointed a town constable, D. W. Howard, but he could not tame the wild elements still left in the small town. In September a killing involving a prospecting group, which included Elko County Surveyor Thompson, Pat Kelley, Walter Waters, and William McLaughlin, took place in Pioneer Gulch. Waters returned from Deeth with supplies for the party, including whiskey. Soon, he and Kelley were "two sheets to the wind." A fight ensued and Waters killed Kelley. Sherwood brought both Waters and McLaughlin to Elko. McLaughlin was later discharged, but Waters had to endure a trial in Elko before he was found innocent because he had acted in self-defense.

In July 1889 Walter Martin, son of Hugh Martin of Island Mountain, captured Tuscarora Jake—an Indian wanderer who had killed two men that April—at the Earl Cassidy Ranch. Cassidy persuaded Martin to postpone arresting Jake for one day so Jake could help finish harvesting the hay crop. Jake was arrested the next day and was brought to Tuscarora and then Elko. Sentenced to the state prison, he died there in January 1892.

Despite its reorganization, the district saw little activity until the late 1880s. By 1890 its population had grown to forty-one, but the Bayard post office

One of the Prunty homesteads at Charleston. (Photo by Shawn Hall)

closed on February 2, 1889. Nevertheless, the town was slowly reviving. It was renamed Charleston after Tom Charles, a local miner. A number of businesses sprang up in the town. The Charleston post office, with Charles Pearson as postmaster, opened on January 31, 1895. Pearson also served as the new justice of the peace. In 1894 Hall, Waters, Charles, and Hamill built a ten-stamp gold quartz mill on Copper Creek and formed the Octorora Gold Mining Company. The Union Gulch placer, which William McCullough and William McLaughlin owned, began yielding reasonable returns. The Bruneau Mining Company formed and local prospectors purchased most of the placer claims. Names of the prospectors and amounts paid were: William McCullough, $6,000; William McLaughlin, $2,000; Walter Waters, $1,000; Ira Bandness, $600; and Mike Pavlak, $150. Pavlak would later be one of the discoverers of Jarbidge and the founder of Pavlak, the first settlement in Jarbidge Canyon.

Production in Charleston continued to be erratic, however. The Octorora Company ran into financial problems, and the property was slated to be sold by the sheriff at Bradley's store in Charleston. The company's last-minute payments of some debts averted the sale, but this was only a stopgap solution. The company folded a couple of years later and the mill, dismantled in June 1900, moved to Tuscarora. Charleston's name had changed but its reputation had not. Violence was common and the sound of gunshots was an everyday occurrence. In fact, lawmen were not particularly anxious to go through Charleston. One deputy returned to Elko pleased but empty-handed after chasing a criminal, saying that he made it "plumb through Charleston and did not get shot."[1]

In 1897 a stage line from Deeth started up with a fare to Charleston of $3.50. The same year, the Rescue mine was discovered. Park City men ran the Rescue Mining Company. They chose Thomas Kearns, principal owner of the famous Silver King mine at Park City, to develop the property. In April the *Gold Creek News* reported that three mills were under construction. Chamberlain and McIntyre built a mill for the Rescue Mining Company, J. A. McBride (first mayor of Elko) built one for S. P. Carlson, and Thuro and Pavlak constructed another at their claims. However, these mills were not very successful, and mining was limited until 1900, when the most prosperous period of production began. The population of Charleston was still at forty in 1900, thanks to the arrival of the Pinkard Prunty family with its twelve children. Demand necessitated the organization of a school. Prunty was named clerk of the school district, and Mamie Dopson was schoolteacher.

A mining revival began in June 1901 when the King Solomon Mining and Milling Company was incorporated. A sawmill began operating, and construction started on a mill located on Deer Creek. The company bought the Carlson, the Prunty, and the Shively mines, and wild claims were published that the mines could produce one million dollars worth of ore. However, the company folded shortly after completing its mill, and in February 1903 S. P. Carlson bought the mill and moved it to his mine.

Despite inconsistent production, Charleston continued to grow slowly. Butler and Cox built the Dew Drop Inn, and a new school, named Bryan, was built. August Keefer ran a fast freight from Deeth and opened a general merchandise store. Charles Lewis also drove a four-horse Concord stagecoach twice a week from Deeth.

In May 1904 the New York and Nevada Mining Company purchased the Carlson mine and mill, and by April 1905 the five-stamp mill was up and running. While some other companies operated in the area until 1910, various members of the Prunty family did virtually all the mining near Charleston for the next forty years. In May 1907 P. R. Prunty completed a small mill at his mine. By 1908 the New York Company had finished a 1,500-foot tunnel on the Black Warrior mine. The Graham (later Slattery) mine, which Charles and Edgar Graham owned, had 1,600 feet of workings. Copper Mountain was the site of most of the mines and a number of them were bonded to the Copper Mountain Mines Company.

The Jarbidge boom gave Charleston a big boost. In August 1911 Miller and Castley set up the Charleston-Jarbidge stage line. Deeth was the best shipping point to Jarbidge, and stage and miner traffic through Charleston was incredible at times. In 1911 the Jarbidge-Badger Mining Company formed, and it worked claims on Badger Creek, north of Charleston. A five-stamp mill began operations in late 1911. In 1912 the Marion Mining and Milling Company came to the district, and the following year a five-stamp amalgamation and concentration mill began production. However, neither mill was a success, and both shut down in 1914. In November 1915 the North Star Mining Company moved the Badger mill to Jarbidge.

Fourteen of the eighteen students at the school at Charleston were Pruntys. In 1915 residents hired a teacher from Ohio, and she soon became Mrs. Mack Prunty. Another teacher from Indiana was quickly hired, and she became Mrs. Ping Prunty Jr. just as quickly. In 1917 a flu epidemic struck, and it killed three Prunty girls and two grandchildren.

While mining continued, the Charleston area was becoming cattle country. The Prunty family had purchased and consolidated a number of ranches. When J. M. Prunty bought John Parkinson's ranch and stock in April 1916, the Pruntys more or less controlled the whole valley. By 1925 they had six ranches along the Bruneau River, a post office on 76 Creek run by an old bootlegger, and the first automobile in the area.

P. R. Prunty continued prospecting on Copper Mountain, and in 1923 the old Prunty mill moved to a lease there, but it met with little success. In July 1926 Prunty leased the Black Warrior mine to Howe and Chamberlain. The two worked the mine for a number of years and in May 1931 completed construction of a ten-ton, five-stamp cyanide plant. It ran until 1937. Some limited mine production continued in the Charleston area until 1942, when Howe and Chamberlain gave the lease up.

The post office closed on July 31, 1951. Over the years, postmasters were Charles Pearson, Reinhold Keefer, Annie Madden, John Prunty, Andrew Knopf, Winnie Shively, Betty Bear, and Stephen Perkins. In 1955 the Pruntys organized the Diamond A Rodeo Livestock Company and ran a successful rodeo circuit. Their claim to fame was the excellent rodeo stock they raised. For years, big rodeos sought their top-of-the-line bulls, bucking horses, and roping stock. The Batholith (formerly Mission Cross) mine was the last to see real mining activity in the area. It produced 400 units of tungsten from 1954 to 1956. Since then, little besides prospecting and drilling has taken place in Charleston.

Today, ranching still plays a prominent role in the area. The Prunty family continues to run ranches there. There are fascinating abandoned homesteads scattered around the valley, and the old schoolhouse still stands, its slate boards intact. However, the town of Charleston has completely disappeared. Some of the buildings were moved to nearby ranches and others succumbed to the elements. The once bustling townsite is marked only by a couple of foundations. As recently as the early 1970s a few buildings still remained. Evidence of mining abounds in the area, particularly north of Charleston along 76 Creek. About three miles north, mine dumps, shafts, and tunnels mark the Prunty (the Graham), the Slattery, the Rescue, and the Black Warrior. The Prunty mill stands near the Charleston townsite. About four and a half miles towards Jarbidge are the remnants of the Mission Cross, or Batholith mine. A small camp formed there, but all the buildings have been removed, and only foundations remain. A trip to Charleston is not only interesting in a historical sense, but the scenery on the way to Jarbidge is some of the most spectacular in the state.

Coal Canyon

DIRECTIONS: *From Dinner Station, head north on Nevada 205 for 1 mile. Exit right and follow the road for nine miles. When the road ends, turn left and follow this road for two miles. Turn right and continue 1 mile to Coal Canyon.*

As early as 1876, notations on maps mentioned a coal deposit in Coal Canyon, but actual development did not take place there until Manuel Sellas and Martin Arrestoy discovered a lead deposit in the area in 1932. The Garamendi mine became the district's primary producer, and in 1933 yielded 49,000 pounds of lead and 135 pounds of copper. However, that was the highest yearly production that Coal Canyon ever saw. In 1936 the Coal Canyon Mining Company took over the property. The following year

the discovery of a copper ore vein at the 500-foot level raised hopes. Joe Riffe began work on a tunnel known as the Coal Canyon mine, which contained lead with small amounts of silver and gold. However, values were not rich enough to warrant continued mining, and all activity had ceased by 1940. The original coal deposit that gave the canyon its name was of poor quality and was never mined.

The Garamendi mine was reworked to a minor extent in the 1970s, but production amounted to only 500 tons of lead ore. Texas Gulf Western did some exploratory drilling in Coal Canyon in the early 1980s, but has done no further work since then. Little besides shafts, tunnels, and mine dumps mark the site today.

Elk Mountain

DIRECTIONS: *From Contact, head south on U.S. 93 for 14 miles. Exit right and follow the road for 5 miles. Turn left and follow the road for 15 miles. Turn right and continue for 7½ miles. Take a left and follow the road for 6 miles. Turn left and proceed 1 mile to Elk Mountain.*

G. W. Gill, J. B. Irvin, and L. B. Foster made the first discoveries on Elk Mountain in 1882, but the Empire State mine that they worked did not prove profitable, and they abandoned it in 1883. W. H. Austeon, a pioneer of Silver City, Idaho, discovered the Austeon Tunnels in 1890, and a number of prospect mines, the largest of which was 200 feet long, started development.

Ore brought three dollars a ton in gold and showed 3 percent of copper, but in what amount is not known, because there is no record of the mines' production. Austeon left the district after a few years.

After the turn of the century, prospectors returned to the area and several new mines started up. They were the O'Neil, the Red Elephant, the Gold, the Estes, and the Robinette. These mines produced little, however, and most never advanced beyond the prospect stage. The district's only period of production took place from 1954 to 1958, when the Pyramid mine, run by the Pyramid Mining Company, yielded a small amount of tungsten. A mill, the Robinette, began operating in 1957, but the company folded the following year.

Since then, the Bunker Hill Mining Company and the Minerals Division of the Union Pacific Railroad Company has done limited mining exploration, but no other activity has taken place. Today, the mill still stands. It looks as if it could start up at any time. Mine dumps, shafts, and dump trucks are nearby. However, the mill is located on a new U.S. Forest Service wilderness area and is slated for demolition. It is terribly unfortunate that many buildings that are critically important to Elko County's history are being destroyed, because they are suddenly part of newly established wilderness areas. It seems logical that these vestiges of the past should be preserved so that future generations have the opportunity to explore and enjoy them. While the ideal of wilderness areas is noble, the history of a place is also vital and should be allowed to coexist with wilderness.

Gold Creek
(Island Mountain) (Penrod) (Goldfield)
(Wyoming District)

DIRECTIONS: *From Wildhorse, head north on Nevada 225 for 2 miles. Exit right and follow the road for 6 miles to Gold Creek.*

First discoveries at Island Mountain were made in 1869, and the Wyoming Mining District was organized on November 3. At the time, the small town of Bruno City was growing just a few miles to the northeast. Early mines in the district included the Mountain King, the Chrysopolis, and the Miners Delight. Additional rich placer discoveries, which Emmanuel Penrod, C.T. Russell, and W.D. Newton made in August 1873, led to the organization of the Island Mountain Mining District on October 11, 1873. Penrod was president and Walter Stofiel was secretary. The placer discoveries led to a flurry of water ditch building throughout the Island Mountain District, and a small camp developed. It was initially named Penrod, in honor of its founder, one of the original discoverers of the Comstock Lode in Virginia City. Most of the people at Penrod were Chinese. The Chinese tended to be more enthu-

The town of Gold Creek in 1897. (Northeastern Nevada Museum)

Gold Creek had a large Chinatown for many years. This is China Lem, a long-time store keeper, in 1910. (Chester and Lillian Laing collection, Northeastern Nevada Museum)

siastic about placer mining and were stoic about hauling heavy bags of placer sand to a remote site where water was available. In June 1874 the Owyhee Water and Gravel Company began hydraulic operations after the completion of a large reservoir and five miles of ditch. The company also constructed a similar setup along Martin Creek at Bruno City. Soon, James Duncan, A.D. Meacham, and Samuel Stanhope were working other placer mines. In June 1875 the camp reached a peak population of 103. Another small camp, called Goldfield, formed near the placer operations on Slate Creek, but these placers were worked out quickly and the camp disappeared.

The main placer area in the district was Hope Gulch, and many companies

worked the gulch. They were the Owyhee Mining Company, which employed twenty; Penrod, Mayon, and Co., which employed twenty and produced $3,000 worth of gold in six weeks; the Kelly Company, which employed five; the French and Green Company, which employed six; the H. W. Brown Company, which employed ten; the William Kidd, which employed five; and Hoover, White, and Company, which employed ten. In July Stanhope and Company sold the Comet mine for $10,000, but the absence of a consistent water supply for placering stifled the camp's growth.

In 1876 Walter Stofiel returned to the district. He had originally come to Elko with a cattle drive, sold his horse and saddle to buy mining equipment, and later opened a store at Penrod. During the late 1870s a Chinatown had developed above Penrod at the mouths of Patterson Gulch and Coleman Canyon. In 1878 Hung Li, formerly from Tuscarora, opened a general store in Chinatown, which he ran through good times and bad until 1918. The Baker school, housed in a small one-room cabin near the Baker home also opened in 1878.

By 1880 Gold Creek's population had dropped to seventy-one, of which fifty-four were Chinese. Stofiel closed his store and moved with his family to Bellevue, Idaho. He later returned and opened another store, with hot springs baths, at Wildhorse. The 1880s were a slow period for the district. Despite this, Ben Yates hauled a ten-ton, water-powered roller mill from San Francisco and installed it at the Cullen mine. Because of Island Mountain's size, a voting precinct formed. Emmanuel Penrod was the voter registry agent. The Island Mountain post office was open from May 5, 1884, to February 28, 1887, and Emmanuel Penrod and Hugh Martin served as postmasters. Penrod had first come to Gold Creek in 1873. Between his placer operations at Island Mountain and nearby Bruno City, he is estimated to have removed $250,000 worth of placer gold. In 1892 Penrod finally sold his extensive holdings in the district to W. B. Duval. He was part of a Colorado syndicate which formed the Island Mountain Mining, Milling, Land, and Investment Company, of which Henry Mayham was president. By April 1893 it employed twenty men. Penrod remained in the area and began work on some other claims in Tennessee Gulch, about five miles to the north, where he joined Walter Stofiel and discovered the Constitution mine. In May 1898 Penrod started up a three-stamp mill, which produced a $1,000 bar of gold in its first nine days.

In 1894 the Gold Creek Mining Company (J. L. Robertson, president) formed, bought out the Island Mountain Company, built the Sunflower Reservoir in 1895, and constructed a twelve-mile water ditch. The company began an intensive promotional campaign and laid out a townsite just to the east of the original town. The first substantial building constructed contained a large hotel, a store, and a saloon—all of which Jack Hardman ran. In 1896 the Gold Creek Company built the Waldron block, part of which contained the Gold Creek Hotel, a 50- by 100-foot, three-story structure that John Abel ran. Smith Van Dreillen started the Elko-Gold Creek stage line, which ran three times a week with a fare of six dollars from Gold Creek to Elko.

Mining opportunities continued to increase through 1896. By November 150 men were working on the placer operations. In addition, four lode mines, the Oro Grande, the Mother Lode, the Blue Ridge, and the Snowflake, were active. In anticipation of future riches, Penrod built a vault to house the gold until it was shipped. Within months, many other new mines were being worked. They were the Diana, the Gold Bug, the Last Chance, the Star, the Constitution, the Empire, the May D, the Third Team, and the Little Maid. On Christmas Eve 1895 the *Gold Creek News* began publication. The paper was essentially a promotional vehicle for Gold Creek. The Gold Creek News Publishing Company owned it; Phil Triplett was the publisher, and Charles Sain was the editor. Sain took advantage of the opportunity to run a serial of his new novel, *Tom Davis in Heaven.*

By January 1897 the boom was on. The town's population was 500, and 300 people were on mine payrolls. The Gold Creek Company employed 240 of these, paying three dollars a day for a ten-hour shift. A telephone line to Elko was completed in June, and the Tuscarora Ladies' Orchestra performed many times at the Gold Creek Hotel. New mines continued to come into production. One was the Star Placers, which a prospector discovered when he noticed a gopher had dug up yellow dirt. Its first 100 tons returned $3,500. In February a shipment from the Little Maid mine returned $277 a ton. The Coleman Placer Mining Company owned the mine, which was the only one in the district with a shafthouse.

A post office opened on February 27 with John Abel as postmaster. Gold

Creek's businesses included the Gold Creek Lumber Company (R. R. Bowles, manager), Anderson and Newhall (assayers), the Gold Creek Meat Market, the Island Mountain Saloon, and J.J. Davis (blacksmith). Clay Irland bought three Mountain City buildings and moved them to Gold Creek, and Dunbar Hunt formed the Gold Creek Townsite Company, which did a brisk business during the spring and summer. Two new stage lines started running when the Dorsey brothers and Fred Wilson began to challenge Van Dreillen's earlier monopoly. Van Dreillen had the semi-weekly mail contract to Stofiel's and later bought out Wilson, combined the routes, and created a daily stage. A big stage war ensued between Van Dreillen and the Dorsey brothers, and fares dropped to one dollar before the Dorseys could no longer compete.

On May 21 the Dwight Babcock family had a boy, the first child born in Gold Creek. The town's first death, a tragic one, took place in August when Shelley Dunlap, a miner, died while working with his partner, Ed Waters. Waters was lowering a bucket of tools when the rope broke and the bucket hit Dunlap on the head, killing him instantly. He was buried in the Gold Creek Cemetery, next to three graves from the 1870s.

In June the Gold Creek Mining Company purchased the old Coptis mine boiler and engine in Tuscarora and moved it to its operations. In July the company patented 4,480 acres of placer ground for $11,200, but the lack of water hindered progress. Despite completion of the reservoir, the ditch was only inching forward, and by August, when the company's property was attached because of a $3,800 debt, hints of trouble were beginning to emerge. While the dispute was quickly resolved, it was a sign of further problems to come. By the end of the year, the situation had become critical for the biggest employer in the district, and in December all the superintendents were replaced. At this point, $600,000 had been spent in development, but the ditch, which had promised to provide an endless supply of water, was abandoned after great expenditures had been made. Compounded with very poor company management, circumstances led to a desperate financial situation. The companies had not paid employees since September. This had a profound effect on the *Gold Creek News,* whose parent company received its publishing funds from the mining company. The paper folded on December 10. Because of lack of money, the last issue was published on the back of a photo. Van Dreillen also pulled out of Gold Creek in December, leaving only a twice-a-week stage from Halleck to serve the community.

The end of Gold Creek was approaching, and while residents enjoyed some rare amenities such as electric lights, a town water system, and telephones, these conveniences did not mean much if there was not any work. The Gold Creek Mining Company closed down its operations in June 1898, while W. A. Smith bought the Gold Creek telephone line for $1,700 in July to satisfy a court judgment. The town of Gold Creek began to empty quickly. By 1899 the Coleman Company was the only major company still active in the area, even though Burkhart and Stauts were sending ore from the Gold Bug mine to Penrod's mill in Tennessee Gulch, which was also working ore from Penrod's own Oro Grande mine. However, the mill and mines shut down in 1900. In a last-ditch effort, the Gold Creek Company named Walter Stofiel manager and superintendent of its closed operations, but the attempt was futile and the Gold Creek Company sold the property to the Coleman Company in January 1900 to satisfy a $17,000 debt. The Gold Creek Company sued, but the Coleman Company still received full right and title.

Gold Creek was dead. The following year, the impressive Gold Creek Hotel was dismantled and moved to Mountain City, where it became the Mountain City Hotel. After 1902, building after building was moved to other locations, and soon the Hardman Hotel, which still housed the post office and a store, was the only business left. Local ranchers Warren Williams, George Williams, and W.T. Jenkins were its main supporters. A fire in September 1921, which started in the George Williams home, burned most of the remaining buildings except the Hardman Hotel. Henry Moffatt bought the hotel and other remaining buildings in January 1929, and the post office closed on February 15. Postmaster Mary Hardman had served since 1897. Immediately after the closure, Moffatt removed all of Gold Creek's buildings, leaving only a concrete sidewalk where a thriving town once stood.

Della Johns, who as a young child helped her parents run the Hardmans' place during the winters when they were in San Francisco, remembers that the hotel had beautiful doors with crystal door knobs, handsome interior woodwork, and fancy furniture. The Hardmans had acquired many expensive collectibles, including two cast figurines which Mrs. Hardman called "Pinky" and "Blue Boy." After the hotel had been torn down, Johns visited the town and found Blue Boy buried in the rubble. She treasures the keepsake of days gone by.

Since 1900 only sporadic mining has occurred at Gold Creek, and its success has been limited. Currently, a small company is working in the old Island Mountain townsite, but although the old vault still stands empty, overlooking the old town, the site is off limits. Higher up in the canyon are remnants of the Chinatown, which contains about ten dugouts. Nearby, a number of old buildings, stone mill foundations, and other ruins remain. The ill-fated ditch is visible to the point where it stops, a scant half a mile from completion. The only reminders of the modern Gold Creek site are a historical marker and a concrete sidewalk. The cemetery, where only Dunlap's headstone is left, is located just to the west. It is believed that George Washington Mardis was buried in the cemetery, after he was murdered near Gold Creek in 1880. His could be one of the three unmarked graves dating back to before the 1890s revival. But the grave's actual location is a mystery.

Hicks District

DIRECTIONS: From Gold Creek, head north for 14 miles. Turn left and follow the road for 5 miles. Turn right and follow that road for 2 miles to McDonald Creek. Hick's Creek is 1½ miles to the north.

The Hicks District was named for "Cap" Hicks, a Cherokee scout who lived in the area for years. Hicks's wife and her family were Shoshones. Her father was the legendary medicine man Chick Chop. A renegade Indian killed Chick Chop's wife, Susannah, on Meadow Creek in November 1882. Chick Chop went to his friend Cap for help. Cap wrote a letter to Jim Dearing, the Indian agent at Duck Valley:

On the night of the 8th this month, "Chick Chop" came to my place at midnight and stated to me that, Yah Un Dicer, and his son, Pime, had killed his squaw and burnt his lodge and also took his ponies, six in number. After resting one day, I sent him back on snow shoes with the understanding he should return if he found things as he represented and proceed to Duck Valley and inform the Indians of the murder.

I waited until the 18th for him to return. I then went down on snow-shoes and found his squaw dead and the lodge burned. The provisions were also taken and their saddles were missing. What has become of Chick Chop is a mystery to me. If he undertook to go to the Martins, he would likely perish before he could reach there. If he undertook to follow the murderers to recover his ponies and he succeeded in over-taking them, they have killed him. There has been no one from the Martins since the occurrence [sic]. If he has reached the reservation you would likely know of it.

As it was one of the coldest blooded murder and robberies that has taken place among the Indians, prompts me to inform you of the facts. Respectfully requesting you to have the Indians apprehended before they do any more mischief as they stated to Chick Chop they intended to go to Rofs place to get his sugar and money. It is my belief that they will murder the first white man they meet. My knowledge of the Indians for the last fifteen years prompt me in making the assertion. They are at present at the hot spring on Bruno where the Indians wintered their cattle three years ago. The weather has been so severe they could not proceed on to Fort Hall as were their plan before they committed the murder. He threatened to kill Rofs with his rifle as he killed the big bears and is not afraid.

I would have went and seen you if I was not so lame from walking over yesterday. They killed a number of Murphy's saddle horses and brought their tails into their camp, which I saw myself. I believe for the interest of our community that the Indian Police should attend to this matter immediately. If Indian Charly, who has two of Chick Chop's daughters for wives, or any other reliable Indian that will come up with others, I will give them all the details and render all the assistance in my power.[2]

Ore from the McDonald mine, discovered in 1872, returned 75 to 100 ounces of silver per ton, but it was extremely difficult to smelt. In 1872 some placer deposits were also found along McDonald Creek, but because of the remoteness of the mines, development was slow and only limited production occurred. By 1874 activity in the area had increased and the Red Cloud, the Congress, the Constitution, the North Star, the Bamboo, and the Bullion mines were being worked. Plans to build a mill in Adder Gulch led to work starting on a water ditch from Hick's Creek to the mill site, but as the work progressed the value of the ore did not warrant the expense of building a mill, and the plans fell through. In August 1876 the McDonald mine was bonded to J.R. Bradley, George Russell, Frank Campbell, and Amos Plummer. The four men did some work on the mine but finally gave up on it in 1879. Henry Catlin leased the mine, and a new partner, J.W. McDonnell, joined him. McDonnell had been in the district for some time and had discovered the nearby Telephone mine in 1878. Michael Enright also entered the Hicks District in 1878 and established the Silversides, the Telegraph, the Fraction, and the Frisco mines. There were eighteen men working various mines and claims in the district by the end of 1880. They included Cap Hicks and his sons, Jess and Eph; Henry Tonkins; William Austin; Ed Youngblood; John James; and Joseph Pearce.

The McDonald mine, under Catlin's guidance, was shipping ore to the Navajo mill in Tuscarora, but even that mill had trouble processing it. Catlin found a smelter in Salt Lake City that could treat the ore, and in September he shipped 22,000 pounds of silver ore there. They returned 100 ounces of silver to the ton. The high cost of production made further work economically unfeasible for Catlin, and he returned the mine to McDonald in 1885. In 1886 he moved his wife and son to the mine, built a home, and continued to work the mine until 1902. In July 1885 Henry Fruederthal and John Carroll discovered the Slate King mine. Other mines worked to a limited extent during the 1880s and included the Herculaneum, the Federal, the Ann Arbor, the Southern Belle, the Evening Star, the Northern Light, the Great Eastern, and the Chipmunk. Cap Hicks and his sons also continued working their claims through 1895. In 1892 Mike Enright discovered ore on his Crescent claims, located on Enright Hill, but these operations made their owners little more than a living, and the men found no valuable veins.

By the turn of the century, only Enright, McDonald, and Jack Chipman were working the district. Enright died in 1902, and J.H. Peck, who did little work, took over his claims. It was not until the summer of 1915, when

Fred Davis paid Peck $20,000 for his claims, that the Hicks District became active again. The following year, Davis dug a 300-foot tunnel, but found little paying ore and left the district in 1917. Little else has occurred in the district since, though in the 1940s the Wicker mine produced some manganese, and in the early 1950s mines on Enright Hill produced some lead, silver, and zinc. Extensive drilling during the 1970s and 1980s revealed some gold deposits, but they were not large enough to warrant mining. Total production for the district stands at 12,700 pounds of lead, 2,943 ounces of silver, 4,900 pounds of zinc, 200 pounds of copper and 16 ounces of gold. Little remains in the Hicks District except some mine dumps and the ruins of a couple of cabins.

Hubbard

DIRECTIONS: *Located 27 miles south of Jackpot on U.S. 93.*

Beginning in 1925, Hubbard was a stop and signal station on the Oregon Shortline Railroad. The stop was named for Smith Hubbard, who had established the adjacent ranch in the 1870s. In 1933 Camp Hubbard, a CCC camp, formed. By 1938 the camp consisted of fifteen buildings, including a large mess hall that seated the 200 men, aged eighteen to twenty-three, who were assigned to the camp. Most had come from a similar camp in Sparta, Illinois. First Lieutenant Edward Clark was the camp's commanding officer, and Dr. Olif Hoffman, also the doctor at the Warm Creek CCC camp, provided medical treatment.

The *Wells Progress* supplied the camp with complimentary papers, and the Camp Hubbard Ranch baseball team, organized in 1933, played against other CCC camps. They played their home games in Wells. While it existed the camp accomplished a great deal, including building and paving large distances of roadway in the region. Hubbard's buildings were removed after the camp closed in the early 1940s, and the Hubbard railroad stop was abandoned when the Oregon Shortline folded in 1978. Foundations mark the railroad stop and CCC camp. Hubbard Ranch, which the Boies family owns, is still operating.

Ivada

(Taylor's Pocket) (Gold Basin Mining District)

DIRECTIONS: *From Rowland, continue north for 2 miles. The road tends to become impassable at this point. Continue another mile to Ivada.*

Located three miles from Rowland, Ivada boomed briefly after the turn of the century. Initially called Taylor's Pocket, promoters renamed the camp Ivada because of its proximity to both Idaho and Nevada. During the summer of 1907 John Blosser and Matt Graham discovered a silver deposit along Taylor Creek. Joe Riffe, Charles Addis, and Ed Blosser soon joined in the venture. Matt Graham, discoverer of the Bull Run mine in the Centennial District, proved to be the "life of the party" at Ivada, where he entertained people by playing the banjo and clog dancing. Ed Hunter was also drawn to the district and staked a couple of claims. The remoteness of the discoveries, however, made growth difficult. Supplies had to be hauled in by mules and, because of the cost, only the highest grade ore was shipped out.

A real estate promoter named Caldwell heard about the potential boom at the tent camp and came to survey a townsite. He laid out 648 lots and sold a number of them but could not provide legal title to the lots because the townsite was on a forest reserve. Caldwell organized the Gold Basin Mining District, for which he acted as district recorder. Supplies for Ivada came from Mountain Home, Idaho, and Elko to the Scott Ranch. From there, pack animals carried them to the camp. By the following year, businesses began opening, including Harry Riddle's general store and saloon, Mulligan Mike's restaurant, M. M. Wolfinger's (Wolfinger was the first woman in town) lodging house, and A. D. Sly's assay office. L. O. Ray, who made his fortune in the early days of Rhyolite (in Nye County), managed the townsite and prospected.

Harry Hunter organized the Gold Basin Mining Company after he paid $30,000 for two claim groups. A number of prospectors formerly involved with the Tonopah boom actively worked claims in the area. They were Jack Flynn, whose mine was the Gold Circles; Harry Campbell, owner of the Heavy Artillery; Kerr and McCoy, who owned the Motherlode; and Envire and Ray, whose mine was the Malmoral. By 1909 the population of the tent town of Ivada had grown to 250. Many years later, Joe Riffe remembered Ivada as "a happy and hilarious place where the inhabitants created their own entertainment," but discoveries at Jarbidge ended Ivada's hopes for permanence.[3] In May 1910 there was a flurry of claim sales in the camp. Hunter sold his claims to John Campbell for $30,000 and George Pitcher sold his five claims for $15,000. Lon Dewey, a millionaire mine operator from Idaho, bought all the other claims in Taylor basin. But it was all a smoke screen. The rush was on to Jarbidge, and the tents at Ivada folded up quickly. By the end of the summer, everyone was gone, and the townsite had vanished into the sagebrush.

Despite the large number of people who had lived there and the amount of prospecting that took place, Ivada produced very little. Most of the activity entailed the selling and reselling of claims, and many who left the camp felt it smelled of a promotional scam. That is probably why Caldwell left soon after he sold lots to which he could not give the owners legal title. Once Ivada was abandoned, only occasional exploration took place there. Since no permanent structures were ever built, not much marks the site except for the leveled tent pads and small dumps scattered throughout Taylor Basin.

Jarbidge

(Jahabich) (Pavlak)

DIRECTIONS: *From Charleston, continue north for 18 miles to Jarbidge.*

Jarbidge was Nevada's last big boomtown, but long before the white man made the canyon prosper, the area was rich with Shoshone legends. For the Shoshone, the canyon was the forbidden home of Tsau-hau-bitts, a man-eating giant that roamed the mountains snatching up helpless Shoshones and returning to the canyon to eat them.

While the discoveries that were to establish Jarbidge did not take place until after the turn of the century, many prospectors unsuccessfully explored the canyon prior to that time. The first group came through in the 1860s and spent a month gold panning the river but found nothing promising. In the 1880s another prospecting party gave rise to one of Jarbidge's most enduring legends—the legend of the lost sheepherder. The story was born when a man named Ross found a rich ledge high on the mountain above the canyon floor. Ross told a sheepherder named Ishman about the discovery and died soon afterwards. While leading his boss, John Pence, back to the canyon Ishman died of a cerebral hemorrhage. Pence continued searching for the mine that, as legend had it, a skeleton guarded, armed with a pick and rifle. Many who later discovered mines claimed that theirs was the Lost Sheepherder mine. Another early prospecting party worked the canyon for placer gold in 1891 and 1892. But, while they found some color, it was not in paying quantities.

David Bourne made the first major discovery in Jarbidge on August 19, 1909, but the Bourne ledge was one of only two ore outcroppings found in the district. Just a few days later, John Escalon, a member of Bourne's prospecting party, found the other outcropping, the Pick and Shovel. While the two mines were close together, the intervening terrain was rough and heavily wooded, and neither of the men knew the other had met with success. Soon

Jarbidge soon after its establishment in 1909. (Ward Morgan collection, Northeastern Nevada Museum)

afterwards, the third member of the prospecting party, Mike Pavlak, made his own discovery, the Pavlak mine. Not only Elko papers but many other publications throughout the West trumpeted the gold discoveries, greatly exaggerating the early value of the mines. One account claimed that $27 million worth of gold was visible in Bourne's mine.

Needless to say, a rush to the area started. The Jarbidge Mining District was organized in October, and at least fifty men lived in tents on the canyon floor, but the approaching winter kept a cap on expansion until the spring of 1910. Harsh winters plagued Jarbidge throughout its history, and the town was often completely isolated for weeks at a time. Until a road was built from Charleston, the only access to Jarbidge was from the north.

The new year started off with a bang when George Winkler, T.B. Beadle, and Edward Benane discovered the Bluster mine on January 11. The Bluster was to become one of the prominent Jarbidge mines. The Jarbidge-Pavlak Mining Company bought Michael Pavlak's mine for $125,000. The Twin Falls-Jarbidge Development Company bought the Pick and Shovel from John Escalon for $250,000 in January, and the Bourne mine now operated under the auspices of the North Star Mining and Milling Company of Boise. In Feb-

The Elkoro Mill.
(Northeastern Nevada
Museum)

The Pavlak Mill.
(Lillian Johns collection,
Northeastern Nevada
Museum)

Ruins of the Bluster Mill, 1993. A major portion of the mill has since collapsed. (Photo by Shawn Hall)

ruary the Jarbidge-Bunker Hill Mining Company organized and bought out the Bunker Hill mines locator, Greenwalt. On March 5 the Jarbidge post office opened, with Roy Nail as postmaster. By April 1, 500 people had flocked to Jarbidge and a city of tents sprang up. The lure of gold even brought Death Valley Scotty, a prominent prospector from the Southwest, to Jarbidge.

By May tri-weekly mail deliveries began from Three Creek, Idaho, and in addition to Jarbidge another small town, Pavlak, began to form near the Pavlak mine. Because of the difficulty of bringing goods in, prices in Jarbidge were extremely high. A hundred pounds of flour sold for twelve dollars, potatoes were fifteen dollars a sack, oats went for twelve dollars per sack, butter was a dollar a pound, canned goods were one dollar each, and a bale of hay cost ten dollars.

Violence made an early appearance in the town. In 1910 Jim Miller killed Bobby Dryne, when Dryne pitched a tent on a lot Miller owned and Miller accused Dryne of trespassing. This incident prompted residents to hire Jarbidge's first lawman, Billy Ross.

Jarbidge continued to grow and develop in 1910, and a small log cabin served as a school. Eleven students attended: Faye McCormick, Marion Strong, Gladys and Herb Pangborn, Pricilla Evans, Lesley Ward, Loren Ray, and Alice, Owen, and Wood Fletcher. Slim Walsh and Vane Whiteside started running pack freight trains from Three Creek, Idaho, and stages, including the Twin Falls-Jarbidge Mining Exchange, which ran a daily from Twin Falls at a cost of thirteen dollars, also started up. The road into Jarbidge was so steep that when a driver approached the town he had to tie a large log to the back of the stage to help slow it down. Log or cut wood structures replaced many tents in Jarbidge during the summer, but as winter approached people left the town in large numbers, not to return until spring. The Jarbidge Commercial Club organized, and charter members elected Judge L. O. Ray president. Tom Beadle donated a lot, on which a twenty-three- by sixty-foot building was completed in August. The club featured reading tables and newspapers, magazines, and mining journals. It also formed a publicity committee to help promote Jarbidge.

By January 1911, 500 people were still living in Jarbidge, which now included twenty-five log houses and seven businesses. In 1911 the town's population rose back up to 1,200, resulting in more modern conveniences, including an $11,000 south road leading to Deeth via Charleston, and a telephone line, also to Deeth, that cost $25,000. Because of increased demand, a new school opened, and the number of students enrolled jumped to twenty.

The biggest obstacle standing in the way of Jarbidge's development was the fact that the entire town was located on forest service land. Because of this, none of the residents could own the land on which their houses and businesses stood. The forest service had to issue special use permits for construction. Because the town was on forest service land, saloons or stores selling liquor were not permitted. But on March 8, 1911, the secretary of agriculture issued a proclamation eliminating the town of Jarbidge from the national forest, thus solving the problem.

In 1911 mining activity occurred throughout the canyon. George Wingfield leased the Bluster mine, while several new mining companies, including the Jarbidge Ely Mining and Development Company, the True Fissure Gold Mining Company, and the Ham and Jarbidge Mining and Milling Company were organized. Two mills were constructed as well. Seventy men worked building the Pavlak mill, a fifty-ton cyanide facility that the Jarbidge-Pavlak Mining Company owned. The mill started up on August 5. Its tower was 109 feet tall, and its top measured 40 by 60 feet. On September 11 the Clark-Fletcher mill, also known as the North Star, started up on Bourne ore. But despite the promising mining activity, there were only a limited number of jobs, and production was minimal and sporadic. By the end of 1911 many of Jarbidge's residents had gone for good, leaving a steady population of about 300.

In January 1912 George Wingfield purchased the Bourne and Success mines for $250,000. All told, Wingfield invested over half a million dollars in Jarbidge's mines, but in March he forfeited virtually all his holdings and left the district. Another blow came in the form of the closure of the Pavlak mill in February. The mill closed because it proved to be too large and was extremely ineffective at treating Jarbidge ore. The Fletcher mill closed the next month after encountering similar problems. Despite these difficulties, Jarbidge had an air of permanence. Businesses included the Mint Bar, the Northern Hotel and Café, W. H. Hudson's general merchandise store, George

Strobel's hardware, J.J. McCormick's merchandise store, and Simon's movie house. The town supported a doctor, an attorney, a dentist, and a reverend. E.A. Hudson also ran a freight line. A short-lived newspaper, the *Jarbidge Miner,* began publishing, and the *Reese River Reveille* wished the paper good luck. That was the last thing anyone ever heard about the publication.

In April 1913 the Potter Palmer Estate, which J.A. Jess represented, purchased control of the Bluster Consolidated Gold Mining Company and bought the Alpha mine for a combined price of $50,000. The Elko Mines Company formed to work the mine, which had been discovered in 1910. A five-stamp, fifty-ton mill at the Alpha mine began operations, but in 1913 the mines produced gold and silver worth only $4,800 in total. The year 1914, during which Jarbidge mines earned $78,000, was the district's first big year. Most of the bullion produced came from the Bluster mine, which now had a ten-stamp mill connected to the mine by a 3,400-foot tramway. The mill processed $70,000 worth of gold and silver in its first six months of operation, despite the fact that ore processing had an extraction rate of only 35 percent. In 1915, thanks again to the Bluster, Jarbidge mines produced $105,000 worth of gold and silver. The Bluster mine was located at the south end of Jarbidge Canyon, close to the small town of Pavlak, whose population rose as the Bluster became increasingly successful. A post office, with James Hayes as postmaster, opened at Pavlak on December 14, 1915, to serve the fifty residents. At this time, a couple of stores and saloons also opened.

In February a strike occurred in the Jarbidge Central mine, which H.O. Milner, Dr. T.O. Boyd, G.D. Aiken, and W.P. Guthrie owned. The North Star Mining Company moved the Badger mill at Charleston in November. In addition, a three-stamp mill began working at the Flaxie mine. By the end of 1915 six mills were running in Jarbidge Canyon. While in 1916 Jarbidge mines produced only a little more than $11,000 worth of silver and gold, the year was one of the more important for the town's future. The Elkoro Mines Company, a subsidiary of the Yukon Gold Company that was associated with the Guggenheim Mining Corporation, formed and would eventually become the most prolific producer of the district. The company spent the year developing the Long Hike and the OK mines. Jarbidge's population still stood at about 300. Because of the large number of single miners, the Jarbidge Bachelor's Club organized.

Many remember 1916 for a tragedy that took place on a dark, snowy night on December 5. Freight driver Fred Searcy was murdered on the outskirts of town while driving his wagon from Three Creek, Idaho. The wagon, along with Searcy's body, ended up on an old side road that branched off the main Jarbidge road. Footprints of a man and large dog led away from the murder scene. While following the tracks, searchers came upon a mail sack. Other mail sacks and almost $3,000 were missing, including $2,700 meant for Newton Crumley Sr., owner of the Success Bar. Sheriff Joe Harris came in from Elko and investigated a 119-foot trail of blood, but the dog tracks were the biggest clue. After the dog was located, he led the group to a bridge, where they discovered a cloth bundle anchored on the bridge's center support containing a black overcoat, a white shirt, a blue bandanna, and a sack of coins. Other evidence included a letter with a bloody palm print on it found in Searcy's wagon. Several residents remembered seeing a local man, Ben Kuhl, wearing the overcoat that had been in the bundle. This led to the arrest of Kuhl and three of his acquaintances. They were Ed Beck, William McGraw, and B.E. Jennings.

The trial began in Elko in September 1917. All four men pleaded innocent. While circumstantial evidence suggested that the others were involved, Kuhl was the focus because one cartridge of a .44 caliber gun he had borrowed from a saloon owner had been fired. But the most convincing evidence was the bloody palm print. An expert testified that the print matched Kuhl's, but the defense contended that the match was just a matter of opinion. The expert proceeded to take several palm prints from court officials and correctly identify every one. The jury only took two hours to convict Kuhl of first degree murder. This case marks the first time that fingerprints or palm prints were used as evidence in a court case anywhere in the world.

On October 18 Kuhl was sentenced to death by firing squad. His execution date was postponed twice and finally commuted to life, after he related a story of how he and Searcy had plotted the robbery, but Searcy did not want to go through with the whole plan. In the course of the argument, Kuhl said, he shot Searcy in self-defense. Kuhl claimed that his accomplices got away with the missing money.

Governor Ted Carville, who had prosecuted Kuhl back in 1917, finally paroled him on April 16, 1945. Kuhl died of tuberculosis a year later. His partner, Ed Beck, went on trial on October 8, 1917, and most of the same witnesses who helped convict Kuhl testified against him. The defense tried to implicate Searcy as an accomplice to steal the mail and money, but the attempt failed. Beck was sentenced to life imprisonment on October 25 but was paroled in November 1923. McGraw and Jennings were eventually released because of lack of hard evidence.

This tribute to Fred Searcy appeared in the *Elko Independent*:

The stage speeds down
Toward Jarbidge town
The driver, his heart is light;
In the iron rack
Lies the money-sack,
With the mail secure and tight,
And he feels content
And confident
That he'll land in town this night.

The road is rough,
And the going is tough
Through the hills that are cold and bare,
But he feels like one
With work well done,
And is glad he will soon be there;
And thoughtlessly
He fails to see
The treacherous, hidden snare.

Foul Murder's head
Rears up so red,
Behind her hideth Greed;
And the crack of a gun
And the deed is done
A soul from a body freed,
And thieves, with ease
The booty seize
And vanish with lightning speed.

And the murdered boy,
The pride and joy
Of his mother, as we all know,
Has paid the price
To Avarice
His life-blood stains the snow,
And the night wind stirs
In the pines and firs
And sobs at the sight below

The hills look down
And darkly frown
On that gruesome, piteous sight
For they feel the breath
Of the angel, Death
And they hear her footstep light
In the twilight dim
As she comes to him
And bears him into the night.

Farewell, Old Pal
So genial
So faithful, tried and true
We will hope and pray
when comes the day
In which we are summoned, too,

At our post we'll be
Found steadfastly,
As staunch to our trust as you.[4]

1917 marked the beginning of Jarbidge's true mining boom. After almost a year of exploration work, the Elkoro Mining Company finally reached a rich ore body in the Long Hike mine and immediately began constructing a 100-ton mill. The company built a seventy-three-mile, 44,000 volt power line at a cost of $250,000. The line led to Thousand Springs, Idaho, and provided power not only for the company's mine and mill, but for the town as well. The company also constructed a dam on Bear Creek and piped water down to supply the town, the mine, and the mills. In addition, a new company, the Long Hike Extension Mines Company, began working the Jarbidge Gold property, just north of Elkoro's holdings. George Wingfield, who could not stay away from Jarbidge, had a big interest in the company. The Elko Prince Mining Company also became active in the district. The Elkoro and the Elko Prince mined about 85 percent of the district's all-time total production. Jarbidge mines produced more than $570,000 worth of gold in 1917, making Jarbidge the largest gold producer in Elko County—a rank it would hold until 1933.

Big things were happening in the town of Jarbidge in 1917. In March investors filed to build a railroad from Twin Falls to Jarbidge, and the Oregon Shortline donated $5,000 towards the project. But building a railroad was unfeasible, and investors dropped the idea in 1918. While the town continued to boom, in 1917 residents experienced a food shortage, because heavy snowfall prevented supplies from arriving. Prices skyrocketed. Potatoes went up to $20 a sack, 100 pounds of sugar sold for $30, and $100 bought a ton of hay.

In March 1918 the Elkoro Mining Company completed the 100-ton Elkoro cyanide plant and mill, located near downtown Jarbidge and connected to the Long Hike mine by a 1,700-foot tramway. The company also began developing the Starlight mine, cutting and hauling timbers for the mines from Deer Creek. Production rose dramatically with the completion of the mill, with $806,000 worth of gold mined in 1918 and another $851,000 worth in 1919, Jarbidge mines' two biggest production years. By June 1919 the Long Hike mine was the largest gold producer in Nevada. But disaster struck Jarbidge on November 3 when a fire broke out. Differing accounts of its origin exist. While one report states that it started in the Jarbidge Commercial Club and spread throughout the business section, another says that a barrel of bootleg whiskey blew up in the Success Bar. The latter account seems more plausible, because there is no mention of the Commercial Club being damaged in the fire. Before the fire ended, fifteen businesses had been destroyed, including the Success Bar, the Mint Bar, the Simon Movie House, the dance hall, the Dozier Restaurant, the telephone office, McCullough's three-story hotel, seven other bars, and five barbershops. While most of these were rebuilt, some of the saloons and hotels never reopened. But the calamity did little to slow down the mines' progress.

By 1920 mining had hit full stride, and there were more than 90,000 feet of underground workings in Jarbidge. Active mines included the Long Hike, the Starlight, the North Star (formerly the Bourne), the Bluster, the Pick and Shovel, the Flaxie, the Legitimate, the OK, and the Alpha. Mills running were the Elkoro, the North Star, the Alpha, the Pavlak, the Pick and Shovel, the Bluster, and the Long Hike. The Elkoro Mines Company was once again the largest gold producer in Nevada. It employed eighty. Wages were $5.50 a day for miners. Muckers, who shoveled ore in the mines, and trammers, who loaded ore buckets onto trams, earned $5 a day. Those who wanted to, could live in company housing for $1.50 a day. The mill's efficiency decreased the cost of processing ore to only $7 a ton. From 1919 to 1921 Jarbidge mines produced a total of $1.5 million worth of gold, of which $1.25 million worth came from the Elkoro-controlled mines. During the late teens and early 1920s the small camp at Pavlak faded as residents moved to Jarbidge, and the Pavlak post office closed on January 15, 1921.

In 1922 the Elko Prince Mining Company left the district after generating $1.3 million in revenue. Besides Elkoro, only the Bluster mine had been a sizable producer. The Elkoro produced $552,000 worth of gold in 1922 and $372,000 worth in 1923. At the Bluster mine, the Bluster Consolidated Gold and Silver Mines Company completed an eighty-ton mill in December 1923. In the spring of 1924 they enlarged it to 100 tons, but it closed in early 1925 when the company went bankrupt. In June the United Eastern Mining Company bought the Bluster, the Pick and Shovel, and the Success mines and refurbished and enlarged the Bluster mill. At the same time, the Elkoro mill was enlarged to 150 tons. Three mines (the Long Hike, the Bluster, and the Alpha) generated revenue in the amount of $571,000 in 1925. In February 1927 the United Eastern Company relinquished its property, and the Bluster mill closed. The Elkoro Mines Company was the only major producer in the district after that. From 1928 to 1931 the Elkoro ranked second in the state, with $2.27 million worth of gold produced. Although in 1932 it produced much less than it had in previous years, the Elkoro still ranked first among Jarbidge mines.

This was the last hurrah for the Jarbidge mines, and only sporadic and minor production occurred afterward. On April 1, 1933, the Elkoro Company shut down all operations, and in September Emil Wolf bought the Bluster property for $50,000 at a foreclosure sale. The Elkoro mill was dismantled in October, and the floor yielded several thousand dollars worth of gold and silver when burned. The dismantling of the mill left the town of Jarbidge without power until the townspeople could raise the $2,000 needed to purchase a transformer. The mining slide continued in 1934, when the North Star Mining and Milling Company sold its property to George Stangn for delinquent taxes. Only the Alpha and OK mines were active. A five-stamp mill ran at the Alpha, and a ten-ton mill ran at the OK. Along with some reworked ore dumps, these two mines, which shut down after 1937, generating $218,000 in revenue 1936 and 1937. In 1937 the Grey Rock Mining Company, a subsidiary of the Newmont Mining Company, took over the Elkoro property. During the next four years, the Grey Rock Company drilled a 1,000-foot tunnel at a cost of $860,000, but a rich ore body was not found. Newmont liquidated the subsidiary in 1941 and left the district. Some leasing activity took place after 1941 and continues today. Some one- or two-man operations have also been around since 1941. A realistic estimate of the Jarbidge mines' total production stands at about $10 million, though some accounts put the amount as high as $50 to $60 million. Materials produced included 355,000 ounces of gold and 1.67 million ounces of silver. Jarbidge ranks first in Elko County silver production and first for gold production in the pre-microscopic gold-mining era.

The town of Jarbidge has survived and is now a popular hunting, fishing, and camping spot. The scenery in and around Jarbidge is a big draw. A gas station, stores, saloons, and a bed and breakfast cater to many tourists. The town hosts a number of popular annual events, including the Memorial Day Barbecue, the July Fourth Barbecue, the Labor Day Corn Feed, and the Halloween Pig Feed. Many original buildings, including the jail and the recently restored Jarbidge Commercial Club, remain amid newer homes. The huge foundations of the Elkoro mill are just off Main Street. The Bluster mill, south of Jarbidge, was virtually complete until partially collapsing a couple of years ago. A cemetery is just north of town. At Pavlak, only foundations and a historic marker show the location of the old townsite. Jarbidge is one of this author's favorite places to visit, not only because it is a wonderful historic site, but also because of the fantastic scenery, which makes it one of the most beautiful areas in Nevada.

Metropolis

DIRECTIONS: *From Wells, head northwest on paved road for 9 miles. For Bishop Creek Dam, head right for 4 miles. For Metropolis, head left for 3 miles. Turn left and continue 1½ miles to Metropolis.*

Metropolis was a planned town, intended to be the center of a huge farming district. The Pacific Reclamation Company and the Metropolis Land Improvement Company organized in 1909. Harvey Pierce of Leominster, Massachusetts, was president of both companies, which purchased 40,000 acres below Emigrant Canyon, including the U-7 Ranch, which Morris Badt owned. Thirty-four homesteads sprang up, and a high-powered promotional

Southern Pacific depot at Metropolis, 1913. (V. W. Birdzell collection, Northeastern Nevada Museum)

The Metropolis Hotel, one of the most expensive and modern facilities of its time. (Mrs. Don Griffeth collection, Northeastern Nevada Museum)

campaign began. The campaign made fantastic claims as to the soil's fertility, but the allegedly rich orchards advertised as being in Metropolis were actually located in Starr Valley. The campaign attracted many young families to Metropolis, but they were not prepared for the hardships to come.

The first phase of development was the construction of the Bishop Creek Dam, which was to provide a plentiful source of water for the town and farms. The scope of the project seems staggering even today. A fifteen-mile road had to be built from the railroad to the dam site. The bulk of the dam

comprised 6.5 million broken bricks left over from San Francisco's devastating earthquake of 1906. Pat Moran was the engineer in charge of the dam, and people still marvel over his accomplishments. A small tent city that included housing for the workers, a store, and a saloon developed at the dam site. The $200,000 dam's head gate closed on April 12, 1911. Laura Sanders of Metropolis christened the new dam with a bottle of champagne.

Once the dam was finished, the promotional campaign became doubly intense. The Mormon church encouraged people to move to Metropolis, and during the summer of 1911 a steady flow of them began to arrive. About 95 percent of the Metropolis population was Mormon. A constant stream of building materials also poured into the town that summer. A townsite was laid out, and construction on many buildings began. The promotion of Metropolis succeeded because 1911 and 1912 were very wet years, and vast fields of grain greeted new arrivals. By the end of 1911 there were 700 people in Metropolis. But the company's claim that thirteen inches of rain would fall each year was entirely untrue.

Metropolis's first major building was an amusement hall that featured a stage and served as the town's social center. Townspeople also used it as a gym, a church, and a theater. The town's crowning glory was the fifty-room, three-story Metropolis Hotel. The brick building cost $100,000 to build. It was extremely ornate, with Douglas fir paneling and a floor made of imported tile. The hotel's first floor also housed a bank, a barbershop, a drugstore, and a general store. The hotel itself offered private baths, electric lights, an elevator, telephone service, and a vacuum-cleaning system. A gala grand opening took place on December 29, 1911, and a special train on the newly completed Southern Pacific Railroad's eight-mile spur from Tulasco to

Metropolis brought dignitaries from all over the West to Metropolis. During the festivities, Lloyd and Lois Dayley were married.

The Metropolis Publishing Company began publication of the *Metropolis Chronicle,* a promotional newspaper, on September 15. The Metropolis Publishing Company was a thinly veiled subsidiary of the Pacific Reclamation Company. The owners sent copies of the paper, which cost two dollars a year, all over the country to attract new residents to Metropolis. A post office opened on November 24, 1911. Jacob Cullen was postmaster.

Metropolis's biggest boom year came in 1912. The town's first school, a two-room stucco facility called Washington Park, opened in January. Claude Schoer was principal. Regular service on the railroad spur, including a daily train to Wells, began on February 18, and a $10,000 railroad depot was completed on March 15. A water system that included fire hydrants ran from Trout Creek to the town of Metropolis in March as well. Pedestrians walked on newly completed concrete sidewalks running between the depot and the hotel, and businesses including several saloons, two stores, the Consolidated Wagon Company, the Stephenson Drugstore, and a fire house opened. A movement to create a new county with Metropolis as the county seat began, and a Latter-day Saints ward formed in February, with Wilfred Hyde as bishop. Metropolis boasted two parks, called Lincoln and Washington, and Metropolis's main streets, Lincoln Avenue and Nevada Avenue, grew crowded with businesses and homes. In addition, a half-mile racetrack near the depot attracted gamblers. The growing town's outward appearance gave every indication that it was on its way to becoming a permanent city.

Metropolis's future looked promising, but already the bottom was beginning to drop out. In June 1912 a group of Lovelock farmers filed suit against the Pacific Reclamation Company, claiming that the Bishop Creek Dam im-

Only the archway of the Lincoln School is left of the once impressive two-story school. (Photo by Shawn Hall)

peded their downstream water rights. This prevented the reservoir from filling, which greatly reduced the number of acres that could be irrigated, and, in turn, forced the Pacific Reclamation Company into receivership in 1913—signaling Metropolis's downslide. The impressive $25,000 Lincoln school was completed in 1913 and had eight classrooms and 150 students. The school printed a newspaper, the *Crimson Clarion,* for many years. The Metropolis Hotel, which the reclamation company owned, closed in 1913. Although the building was still used for dances and weddings, its closure cast a pall over the town. On a more positive note, the Southern Pacific Railroad awarded the Metropolis depot a silver medal for being the "best appearing" station. The depot featured extensive flower gardens and a large fountain, but it could not brighten the dark years heading Metropolis's way. The *Metropolis Chronicle,* with little left to promote, folded on December 15, 1913.

Metropolis's wet years ended in 1914, when exceptional dryness devastated crops. The dry conditions gave rise to two additional problems: Mormon crickets and jackrabbits, which destroyed anything that did manage to grow. The drought extended through 1918, and the crickets and rabbits stayed around as well. The rabbit invasion was so overwhelming that one farmer commented that if you killed one thousand rabbits there would be ten thousand at the funeral. In 1915 and 1916 the combination of drought and pestilence caused the complete failure of eighteen dry farms in the Metropolis area. Thirteen others failed almost entirely, and only a handful managed to survive. In the mid-teens most of Metropolis's businesses and residents left, many heading for Gridley, California. Some hardy farmers tried to stick it out, but few stayed on. To add insult to injury, a typhoid epidemic hit the town in February 1916, killing many people. The epidemic hit the Hyde family particularly hard. Mormon bishop Wilfred Hyde and his two brothers, John and George, all died within three weeks, leaving seventeen children fatherless. One of these children, Carla, was the first girl to have been born in Metropolis in May 1913.

In October 1915 Bond and Goodwin purchased the remnants of the Pacific Reclamation Company and the Metropolis Improvement Company for $200,000, but the pair was more interested in selling the property than in doing anything with it. By 1920 less than 100 people remained in the town and on the surrounding homesteads. The 1920s were very quiet in Metropolis. Little activity took place, and the main sources of entertainment were the school's athletic teams and the local baseball team. The last store in town, which Farren and Jensen ran, was robbed in June 1925 and closed soon afterwards. The Southern Pacific Railroad abandoned its spur in 1925 and tore up the rails in August. The hotel, which had been effectively abandoned, fell victim to vandalism, and the beautiful hardwood floor was destroyed. In September 1929 the amusement hall and meeting house, which also housed the Latter-day Saints church, caught fire and burned to the ground for a complete loss of $4,500. The water system had been abandoned, and there was no water with which to fight the fire.

While some farming was still going on in Metropolis in the 1930s, the town was basically dead. The school still operated, and a few people lived in town, but all outward signs of life except the post office were gone. On September 11, 1936, the hotel, now an empty shell serving as a home for vagrants, caught fire. The blaze began spreading and only the residents' heroic efforts saved the school and gym. The hotel, once the pride of Metropolis, was gone, and the remaining walls were dynamited for safety reasons. The Lincoln school closed for good in 1943 and was dismantled in 1946 for its bricks. Only the large concrete archway remained standing. Postmaster Doris Hutchinson closed the post office doors for the last time on December 10, 1942. The last stronghold of the town, the grammar school, closed on March 1949, when the school consolidated with Wells. The impressive town of Metropolis had became Nevada's only true agricultural ghost town.

Over the years, the buildings in Metropolis either succumbed to the elements or were moved to other locations, but former residents still feel a strong bond with the town. On July 22, 1989, former residents dedicated a memorial plaque at Metropolis. More than 150 original residents and their descendants attended the ceremony. The plaque, with photographs and a map of the town, reads:

REMEMBER METROPOLIS

In memory of those valiant pioneers who settled and built a city here, giving so much to us all in their pursuit of happiness and security. Today we enjoy the fruits of their efforts. The first settlers came in 1910, followed by many others in 1935. Many who lived here aspired to become teachers, lawyers, civic leaders, church leaders, and best of all reared great families in homes where love and happiness filled their lives. May we always remember our Metropolis heritage and beginnings, and may our lives be fuller because of the Metropolis pioneers. May we also forever resolve within our hearts and minds to cherish the memories of the pioneers of Metropolis. Blessed be the name of Metropolis, Nevada through the eternities.

Metropolis residents placed a historical time capsule containing their family histories in the monument.

While the town has been abandoned, the area includes six ranches that evolved from early homesteads. The Bishop Creek Dam still stands proudly, but the creek flows through an open gate. The hot springs, which Metropolis residents enjoyed and which served as the main baptism site for new Latter-day Saint members, still flow below the dam. No buildings are left at the town of Metropolis, but many ruins tell the town's story. The basement and arch of the school are intact, and the bank vault sits in the middle of the huge hotel foundations. Many other foundations show the layout of the town, while large sections of the concrete sidewalks are left. The railroad depot's foundations are located south of town. A small cemetery stands on a hill to the east and contains six Rice family graves, and the extensive Metropolis, or Valley View, Cemetery west of the school offers mute testimony to the hardships Mormon settlers endured. The great promises of what amounted to a promotional scheme drew them to the area, but they managed to scratch a living out of the harsh environment.

Mountain City
(Cope) (Placerville) (Fairweather) (Sooner) (Murray)

DIRECTIONS: *Located 84 miles north of Elko on Nevada 225.*

The first spark of interest in what would later become the Cope District ignited in the fall of 1868, when M.L. Henry explored some placer gold deposits in the area. But Jesse Cope's discoveries in April 1869 are what really led to the Cope boom. Cope's main attractions were the Pioneer and Argenta mines. Within months of their discovery a few hundred people had already come to the area. During those early days, a small fort, called McGinnis, provided protection against Indian attacks that never occurred. The Cope Mining District organized on May 22, and soon afterwards many new mines became active. These included the Mountain City, the Fritz, the Nevada, the Idaho, the Crescent, the Buckeye, the Monitor, the Great Eastern, the Mammoth, the Hamilton, the Idaho, the California, the Sunny Hill, and the Crown Point. A mining district adjacent to the Cope District, called Murray, organized in July. It included the Eclipse, the Wool, the Kansas, the Black Eagle, the Raven, the Grey Eagle, the St. George, the St. Andrews, the Lodi, the Excelsior, and the Eldorado mines.

By June Cope's population stood at 300, and on July 13 the booming camp was officially renamed Mountain City, although many continued to call it Cope. Colonel Frank Denver sent its first major shipment of 300 pounds of high-grade ore to the Miners Foundry in San Francisco, which assayed it at $420 per ton. Argenta mine ore headed to the Dall mill in Washoe County. Frank Drew owned the mine, and after three tons of ore returned 470 ounces of silver, he boasted that the mine was richer than any in White Pine. Other new mines included the Rattlesnake, the Blue Bell, the Fuller, the Estella, the Little Giant, the Virginia, the Tiger, the Keystone, and the Grant. The Rattlesnake mine showed incredible initial values of $9,800 per ton in silver.

By the end of summer Mountain City's population had grown to 700, the town had twenty saloons, a dozen hotels, six restaurants, and two breweries had opened. Four stores also opened; they were Oppenheimer and Company,

Fish and Fellows, Chapman and Blair, and Goff and Company. Construction began on a $10,000 water ditch from the Owyhee River. The ditch was built to help work placer gold deposits. In its first year, two stage and freighting companies served the camp. They were M. P. Freeman and Company's Elko and Mountain City Express and J.T. Enright's Weekly Express. At the same time, Chinese placer miners began moving into the area, and the suburb of Placerville began to develop. But by October the initial excitement had died down, and around two hundred people left the district. The completion of the Elko and Idaho toll road in October made travel to and from Mountain City much easier. The town received a boost in November when the $25,000 Drew and Atchison ten-stamp mill, which contained a thirty-five horsepower engine and four amalgamating pans that could process eighteen tons of ore a day, started up. The Cope Mill and Mining Company was organized and ran the mill and the Argenta mine. This allowed many of the smaller mines, which otherwise could not have afforded to ship ore, to remain active. Two new mines, the Borealis and the Crescent, began production, and Mountain City mines produced $25,000 worth of copper, gold, and silver in their first year.

Mountain City mining boomed in the following year, when mines earned $175,000. Improvements at Mountain City began in 1870. On January 12 the Canty and Allen mill, formerly an exhibit at the Mechanics Fair in San Francisco, started up and began processing ore from the Crescent mine, and a post office, with George Tunney as postmaster, opened on February 12. In April another nearby mining district, the Fairweather, organized. The Sooner

Mountain City boardinghouse, 1903. (Elko Free Press collection, Northeastern Nevada Museum)

Mining District was also located in April and officially organized in July. On April 30 the Reinhart, Wingard, and Drew mill, which the Cope Company owned, started up and worked Argenta ore. By June Mountain City's population stood at 1,200. The town's most productive company was the Argenta Silver Mining Company, which T. J. Kiely owned. He had purchased the Argenta mine from the Cope Company, which generated revenue in the amount of $100,000 in 1869 and 1870. Smith Van Dreillen put a fast freight line into operation, and the Northern Stage Company made Mountain City the terminus for its line, with a fare from Elko of twelve dollars. Stoddard and Thompson built the Argenta Hotel that opened along with the Humboldt Restaurant and Arcade Saloon. There was a big celebration on July 4, and the featured event was a challenge horse race between "Baldy" and "Walla Walla Pony." Baldy won by four and a half feet, winning his owner $300. In October the thirteen-stamp Norton mill, which the Excelsior Mill and Mining Company owned, began operations. The following month, the ten-stamp Vance mill started up. By the end of 1870 Mountain City contained more than 200 buildings and had a population of 1,000. Businesses included twenty saloons, nine stores, six restaurants, two bakeries, four blacksmiths, and two breweries. There were also six lawyers and three doctors in town. Many of the early mines had played out, and five major mines were the focus of operations. They were the Argenta, the Excelsior, the Crescent, the Monitor, and the Buckeye.

By the beginning of 1871 three mills—the Vance, the Norton, and the Drew—were in operation, and in March R. F. Davis bought the Vance mill and added more stamps. At the same time, a smelter, the Robbins, was completed. Two new mines, the Hunter and the Independent, began production, but by summer Mountain City's population had dropped to 450. Principal mines included the Independent, the Argenta, the Pride of the West, the Eldorado, the Monitor, the Excelsior, the Constitution, the Mountain City, the Grant, the Crescent, the Idaho, the Nevada, the Emmett, the Hunter, and the St. Nicholas.

With the big boom period over, Mountain City settled into a steady state of existence. A school opened in July. In November the Excelsior Company completed a twenty-ton Bailey furnace, the first of its type in the West. In 1871 mines produced a little more than $89,000 worth of primarily copper. In January 1872 revenue generated amounted to $18,896, but the mines had begun to slow down. In 1871 they earned only $36,054, and the figure fell to $33,000 in 1873. A big blow came in the form of the closures of the Excelsior mine and the Norton mill. Mountain City's population had shrunk to only sixty-seven by 1875, and in 1876 mines earned only $6,600. Mining revived from 1877 to 1880 when the district produced bullion worth almost $400,000, but quickly thinning ore veins and the plummeting value of silver killed Mountain City. By the end of 1880 its population was only thirty-five, and virtually all its businesses had closed. By 1882 only twenty people were left, and from 1881 to 1904 mines that produced sporadically throughout this period yielded only a little over $43,000 worth of ore.

During these lean years many homesteaders came into the area, and a large ranching industry, which survives today, developed. Even though mining had virtually ceased, the huge tailing piles blocked part of the Owyhee River during a high water year in 1887, forcing migrating salmon to divert up Indian Creek. The Resurrection Mining Company and the Rocket Mining Company were active in the late 1880s, but they produced only insignificant amounts of ore. Clay Irland and Lewis Goldstein hauled Resurrection ore to Elko. Goldstein owned a Mountain City store. The Resurrection Company built a small mill and two concentrators that began operations under John Ainley's guidance in November. Mountain City's school was still open, and in December 1889 Dave Casper purchased Goldstein's general merchandise store. By 1890 the Resurrection Company had folded. Little mining was conducted in the area during the next few years.

Mining activity that resumed in the late 1890s produced the lion's share of the $43,000 worth of copper, gold, and silver mined at Mountain City from

1881 to 1904. The Walker brothers made a rich strike in September 1897. In December E.N. Gray, Frank Ish, and C.A. Watkins incorporated the North Elko Gold Mining Company to work the Rosebud, the Ethel, the Hawk, the North Star, and the Rising Sun mines. Most of the original buildings at Mountain City had been moved or torn down since the boom years, and a small, new town began to develop a short distance from the old Cope townsite. A newspaper, the *Mountain City Times,* began publication on January 21, 1898, using the old *Gold Creek News'* plant. Charles Sain, R.M. Woodward, and Charles Keith owned the four-page paper, which cost five dollars a year. That was an optimistic price, however, and the paper folded on May 13. A board of trade, a type of chamber of commerce, organized in February. Businesses that opened during this mini-revival included J.B. Hall's general merchandise store, Joseph Peck's hotel and saloon, and the Blue Mule Saloon, which Winter and Oldham ran. The Mountain City school was reestablished in 1898, and teacher Lydia Hosking had ten students. Hawes and Poolin instituted the Tuscarora–Mountain City stage line, and the tri-weekly stage cost five dollars from Tuscarora to Mountain City. In April the Walker brothers bought the old Vance mill, which had originally cost $40,000, for $1,250. The brothers refurbished and restarted the mill. In May C.D. Lane bought the Curieux group of mines for $150,000, and by 1900 Mountain City's population had rebounded and now stood at 100.

In September 1904 S. Newhouse paid $40,000 for the Greenback mine, which Butler and Walker had discovered in May. The mine employed fifteen men. The Nelson and the Protection mines also began production. The Protection Mining Company, which built a ten-stamp mill at the mine, was the main producer during this revival, which lasted until 1908. The Nelson Mining Company built its own fifteen-stamp mill, the McDonald, but legal problems that hindered production plagued the company. The Greenback, the Protection, and the Nelson mines generated revenue in the amount of a little less than $27,000, before the revival fizzled in 1908. In 1907 the Protection mine alone was responsible for $18,000 of that amount. Productive mining in Mountain City came to an end in 1908. In 1920 the town once again saw some limited activity when Mark Hopkins bought the Nelson mine. The following year a fifty-ton flotation plant and a ten-stamp mill began operating at the mine. The mine earned $180,000 in 1921, but the veins tapped out, and all operations shut down by 1922. Mining took its last gasp in Mountain City from 1925 to 1930, when the Silver Banner Mining Company worked the Nelson and Mountain City mines. The mill restarted in June 1925 and expanded in 1928, but it earned only a few thousand dollars.

It was not until the great copper discoveries were made at nearby Rio Tinto that the town revived. Businesses still operating in Mountain City relied on Rio Tinto mine employees for support. These businesses included Bill Doyle's Mountain City Hotel, Manuel Bastida's store, the Davidson brothers' store (which Jack and Walt Davidson and Tom Stinson opened in 1919), and the Mountain City Freight Line, run by George Irland. The Mountain City Hotel had been moved from Gold Creek, when that boomtown went bust. On June 30, 1933, the *Mountain City Messenger* began publication. Derf Delanoy owned the paper; the first two issues of which were mailed free of charge to residents in order to solicit subscriptions. The ploy was initially successful, but the two-page paper had folded by May 1934. Improvements to the town included a water and sewer system built by Robert "Doby Doc" Caudill in 1937, though because of the system's problems, many residents were not happy with his efforts. The *Mountain City Mail,* run by Clayton and Clinton Darrah, began publication on May 19, 1938. The weekly soon had 400 subscribers, but the Darrahs suddenly closed it down on February 23, 1939, to run a paper in Reno. In 1945, after owning their store for twenty-six years, the Davidson brothers sold it to H.C. Read, who renamed it the Golden Rule. The Golden Rule chain grew and eventually became a forerunner of the J.C. Penney Company. When the Rio Tinto boom died out in the late 1940s, Mountain City slowly shrank—though it was still home to about 100 people, and a number of businesses continued to serve the residents.

During the 1950s the discovery of uranium on nearby Granite Ridge stirred up some excitement in Mountain City. Milt Ray and Ed Gregory organized the Mountain City Uranium Company in 1956. The Continental Oil Company leased the property in June 1957 but quit a few months later. In March 1958 the Standard Slag Company of Gabbs (in Nye County) leased the property, and in July found molybdenum but not uranium. Despite all the interest, very little uranium ore was ever shipped. The only decent producer of uranium was the Racetrack mine, which yielded 10,000 pounds from 1958 to 1963, making it the largest producer in Elko County. In October 1958 the Bagdanich Development Company began construction of a seventy-five-ton mill, which included an assay office, to treat ore from claims in Gold Creek. George Caron Enterprises bought the mill in June 1963 and redesigned it to produce a soil conditioner. When it was completed, Nevada Secretary of State John Koontz pulled the power switch. However, the ambitious plans did not come to much and the mill was converted again, this time to produce tungsten. The mill closed for good in the 1970s, and the property was dismantled. The U.S. Forest Service reclaimed the leach ponds.

There is much to see in the Mountain City area. At the older Cope townsite, many foundations and pieces of debris are left. Unfortunately, only dugouts and depressions remain at the Chinese camp of Placerville. During the summer of 1992 the U.S. Forest Service extensively excavated two Chinese dugouts in Placerville. At one time, fifty homes stood there, forty of which were owned by Chinese. The camp was across the Owyhee River from Mountain City, not at the present Mountain City townsite, as is stated in some history books. Most of the Chinese residents were involved in gold placer mining, and items retrieved during the excavation provide the most accurate information to date about Chinese homes in Nevada during that period. Remnants of the large water ditch the Chinese constructed from Mill Creek are located nearby. Mountain City remains a quiet little town with a population of about seventy-five. Businesses there rely on tourist and sportsmen traffic. The town's proximity to Wild Horse Reservoir helps keep them afloat. A number of old buildings are interspersed amongst newer structures. Lodging, food, and gas, among other things, are available at Mountain City.

North Fork
(Johnson Station)

DIRECTIONS: *Located 51 miles north of Elko on Nevada 225.*

A settlement at North Fork first began to form in 1870 when William and Catherine Johnson moved from the original Johnson Station, located on Walker Creek, and reestablished the station at North Fork on the newly completed Cope Road. A daily stage came through, and Johnson Station did a thriving business feeding and lodging travelers. The Johnsons had eight children, five of whom were born at the ranch in North Fork. William died suddenly in December 1877 just months before his youngest child, Lillian, was born. A few other families, including the John Walker family, came to the North Fork area in the late 1870s. In 1880 North Fork's population was thirteen. This included Catherine Johnson and her six surviving children, all of whom were under the age of twelve.

A number of ranches began to form, and large cattle outfits such as the Morgan Hill Company and the Murphy Cattle Company used the range. The larger ranches consisted of many smaller ones that were scattered throughout the expansive North Fork Valley. A small town began to form at North Fork, and it became the ranch's social center. George Pratt moved to North Fork, and when his brother-in-law, Richard Morse, joined him later, the two established a freighting business to Tuscarora and Midas. Morse, with his wife and five children, settled on Morse Creek, where he built a sawmill that produced mine timbers for Tuscarora. The Morse Ranch stayed in the family until 1948. By the time a post office opened at North Fork on January 17, 1889, with Dollie Shearer as postmaster, a hotel, a saloon, a grocery store, a stage station, and a school were in operation, and seventy-five people lived in and around North Fork.

In the 1890s Manuel Larios built a new stage station that offered first-class meals, lodging, and a stable. A small store featured wines, liquors, and cigars. In 1897 Henry Van Dreillen began running a stage from Elko through North Fork to Gold Creek. The stage also carried the mail. The fare

Remnants of the stone store at North Fork. (Photo by Shawn Hall)

from Elko to North Fork was four dollars. In March the Bowles lumber mill started to provide lumber for the boomtown of Gold Creek. In June 1900 J. D. Franklin organized the tri-weekly North Fork–Gold Creek stage line. North Fork was home base for the line, which connected with the Tuscarora/Elko stage at Dinner Station. Because of the extensive ranching industry in North Fork Valley, the North Fork Cattle Association formed. They held meetings at Domingo Hall at North Fork. In 1900 the valley's population stood at 122.

At their ranch located across the road from the businesses at North Fork, the Johnsons built a new two-story frame home. Up to the time they completed the house in 1901, the family lived in the log buildings erected in the 1870s. In 1906 Bill Mahoney built a large stone building that housed a store, a saloon, and the stage stop. The North Fork post office was located in the new Johnson home, which also served as the lunch stop for the Elko–Mountain City stage. After Catherine Johnson died in 1908, two of her children, Lillian and Emery, took over the ranch and other operations. Lillian became the North Fork postmaster in 1909, a position she held for thirty-five years until the post office closed on June 30, 1944. In 1912 Lillian married Chester Laing and the couple bought Emery out. Percy Royals and Laing then ran the store and built a dance hall next door. Classes at the North Fork school, sometimes called the Harrison school, initially took place in a log cabin on the Johnson Ranch. Later on, the school moved to a room in the new Johnson ranch house. The original school subsequently became part of the Robert "Doby Doc" Caudill collection featured in the Lost Frontier Village in Las Vegas. The whereabouts of the school is a mystery today, since it was sold after Caudill's death. Two additional schools served other parts of the valley. The North

Humboldt school, where Flo Reed received $120 a month to teach seven students, was located at the McKnight Ranch during the 1920s. The school was a twelve- by twelve-foot adobe building with one-inch thick walls. In 1932 the Mahala school opened at the Tremewan Ranch, where it remained for a number of years.

As time passed the stage lines stopped running, but the store and the saloon were able to survive because of automobile traffic. In the 1930s Chester and Lillian Laing built a new store, the North Fork Mercantile, next to their ranch. In August 1934 Wallace Frost and Earl McCullough robbed the store of clothes, whiskey, bacon and saddles, worth hundreds of dollars altogether. The men were later arrested. The store, which was North Fork's last business, burnt to the ground and was never rebuilt.

Ranching continued to play a big part in valley life. In the early 1940s Newt Crumley Sr. bought many of the valley's ranches, including the Kearns, the Evans, the Bellinger, the Saval, the Truett, and the Tremewan. In July 1947 he sold all of them except the Saval to Bing Crosby, Elko's new honorary mayor. Crosby combined all the ranches to make a single large one, called the Bing Crosby or PX Ranch. He spent as much time as he could there, and his family lived there as well. John and Doris Eacret managed the ranch. In June 1948 Crosby also purchased the Johnson, or the Laing, Ranch from Chester and Lillian Laing, who had retired to Elko. In September of the following year Crumley sold the Saval Ranch and retired from the cattle business. In December 1953 a fire destroyed the Johnson Ranch house.

Crosby ran the PX Ranch until November 1958, when he sold all his holdings in North Fork to Edward and William Johnson and Earl Presnell for more than $1 million. While ranches in the area have changed hands over the years, North Fork continues to be a prosperous ranching district. At North Fork, the stone store struggles to remain standing, but the dance hall has collapsed. For years a bottle house stood at North Fork, but it is only a memory today. A huge barn marks the location of the old Johnson Ranch. A number of foundations and other ruins are left at North Fork. They exist side by side with a modern-day Nevada Department of Transportation maintenance station.

Patsville

DIRECTIONS: *Located 2.3 miles south of Mountain City on Nevada 225.*

Patsville came into being as a result of the boom at nearby Rio Tinto in 1932. The small town was named for Pat Maloney, who, along with Marge Clark, ran a sporting house and dance hall. In the 1930s and 1940s Patsville had a drugstore, a service station a garage, a boardinghouse, saloons, and a red-light district. Many miners had homes in Patsville, and the population remained at around fifty throughout most of the town's history. In May 1937 Maloney sold the entire town of Patsville to William Doyle, who formerly ran the Mountain City Hotel, for $15,000. Once the Rio Tinto boom came to an end in 1947, Patsville began a quick slide into oblivion. The town's last store, which Chauncey and Gen Olson ran, closed in 1949. A number of buildings still stand at Patsville.

Rowland
(Bueasta) (Diamond A)

DIRECTIONS: *From Gold Creek, continue north for 18 miles. Turn left and follow the road for 1½ miles to Rowland.*

While a number of ranches were homesteaded in the Rowland area in the 1880s, the small ranching community did not really form there until the 1890s. For years, cattle outfits had used the nearby Diamond A range. At one time Dan Murphy—a member of the Murphy-Stevens-Townsend party— the first wagon train to reach California in 1844, had 20,000 head of cattle with the Diamond A brand. The first post office in the area opened at the homestead of Joe Taylor, the first settler in the area. Taylor garnered 100 petition signatures to establish the office and arranged for the mail service to run from Gold Creek to the Bueasta office, which opened on November 17, 1896. Taylor was the postmaster.

The name Bueasta came from a combination of the names of three local ranchers, Frank Buschaizzo, Lou Eastman, and Joe Taylor. During the Gold Creek rush, the Mountain Home–Gold Creek stage ran through the Taylor Ranch. Other ranchers around Rowland included William "Red" Strickland, Fred Anderson, Ed George, Hiram and Bert Salls, Don Williams, Sam Bieroth, and George Rizzi. Taylor's wife, Maggie, was a full-blooded Paiute and her mother, Susannah, was married to Chick Chop, a chief. Bannock Indians attacked and killed her while she was camped along Meadow Creek. Maggie later became a Gold Star mother when her son, Willard, was killed in France in Word War I. Two of her grandsons, Elmer Purjue and Albert Arendt, died in World War II. They were the only war casualties from the Rowland area.

Around the time that the Bueasta post office opened, local residents formed the Bryan voting precinct in honor of presidential candidate William Jennings Bryan. A sawmill was active in the town in the 1890s. The Bueasta post office closed on February 15, 1898. A short time later, John Scott bought the Joe Taylor Ranch for $3,500. A new rancher, Rowland Gill, settled in the area. When a new post office opened on March 5, 1900, with Charles Tremewan as postmaster, it was called Rowland in his honor. The office was located in

the Scott house, a large home made of rough logs to which Scott had added a store. Scott was very generous, and everyone who came to his store received credit. The store was vital to the majority of residents, since most winters the roads to Rowland were closed because of snow, and travel was often limited to horseback, sleigh, or skis. Scott's wife was Elizabeth Walther, whose father, Valentine, lived in Sherman, in southern Elko County. Scott, whose last name was originally Scotto, was born in Genoa, Italy, in 1854 and learned English while working at the Carville Ranch.

Rowland's first school was established in the Scott home in 1900, and the nine Scott children made up a large percentage of the student body. A couple of years later a new schoolhouse was built out of lumber. An organ was installed, and the school became the site of popular local dances. In the early 1900s other schools opened at various locations in the Rowland area. One opened on McDonald Creek. The first teacher was Ray Hagenbough— an eccentric English rancher who had graduated from an Ivy League college and was later an ace pilot in World War I. He built a private tennis court at his ranch. Hagenbough and his wife, Winnifred, bought the old Scott Ranch in 1936. The school remained open until 1918. The Rimrock school was built on the Diamond A in 1925 at the Strickland Ranch. The school, which closed in 1933, was accompanied by a separate cabin to house the teacher. One of the teachers, Hedlig Cazier, later married Joe Taylor Jr. Another school operated at the Sam Baker Ranch for a few years in the 1920s. Mary Scott was the teacher. Yet another school was located on Meadow Creek during the 1930s. The school burned in 1938, and a cabin was moved to the site to serve as the replacement school. The schools did not necessarily remain permanently at their original locations but moved from ranch to ranch to serve the family with the most children.

For many years, Rowland offered locals and travelers a post office, a store, an assay office, and a saloon. In 1918 Scott built a new store out of rock. It became the area's main supplier of groceries, clothing, and farm supplies. By 1920 John Scott owned most of the Rowland area, including the successful saloon, which served the area's thirsty cowboys. Beginning in 1924 mining started to have an impact on Rowland, when Roy Cook and Jack Goodwin incorporated the Bruneau Gold Mining Company. Joe Riffe had actually started the mine a number of years before. Riffe had built an arrastra but sold out after making a marginal profit.

Cook and Goodwin immediately made plans to build a mill. They erected a power plant, a dam, and a flume system. By September 1925 they had begun constructing a mill, as well as a bunkhouse and a cookhouse. They completed a new dance hall and store on the property in February 1926. Soon afterwards, they finished work on the three-stamp, twenty-ton amalgamation mill, and the mine had 700 feet of workings. In May 1926 Tony Scott discovered his so-called "Dream" mine, which he found after a dream revealed its location to him. The first rich strike took place at the Bruneau mine in July. It assayed at $103 per ton, but veins in the area tended to be short and small. In August 1927 the Bruneau mine and mill closed because of a lack of funds, and Scott laid off all the workers except security guard Chauncey Olson. Scott sold his now worthless Dream mine for $100. It had produced only 180 ounces of gold and 86 ounces of silver.

In February 1930 Scott, a long-time resident of the area, died. Soon afterwards his wife, Elizabeth, sold his holdings to George Rizzi. The Scott store closed, and soon after Scott's death the development at the mine became Rowland. The post office and school moved to the mine and mill, which were located about a mile south of town. The Blewett brothers purchased and reopened the Bruneau mine and mill in 1930. They hired eleven men and selected the mine's former owner, Roy Cook, as the manager. The brothers added a fifteen-ton cyanide plant to the mill in August. Chauncey Olson built a new store near the mill and also hosted weekly dances. However, production during the following years was limited, and both the mill and mine operated intermittently until 1939, when all mining there ended for good. At the time that mining operations were curtailed, five men—Golden and Ashel Hyde, Verland Stowell, Tony Scott, and Sim Davis—worked at the mine. From 1930 to 1937 the mine yielded 139 ounces of gold and 56 ounces of silver. Olsen's store remained open. In 1936 it survived a fire, thanks to a quickly formed bucket brigade. The store finally closed in 1942, when Olsen sold his ranch and left Rowland. He and his wife, Gen, bought a store at Patsville, near Rio Tinto.

Rowland once again became strictly a ranching community. A major improvement in the lives of residents came in the form of the installation of telephones in their homes in the 1930s. The post office closed on November 14, 1942. Over the years, the many small ranches that made up the Rowland area have been purchased and combined to make larger ones. At "old" Rowland a number of buildings from the early years remain, including ruins of Scott's new store and saloon and the old storehouse, which is now used as a storage building. The ruins of the mill are located at "new" Rowland, to the south. While an official cemetery was never organized, a small graveyard exists at the old Scott Ranch, beside an orchard.

Stofiel
(Butlers) (Wildhorse)

DIRECTIONS: *From Mountain City, head south on Nevada 225 for 20½ miles to Stofiel.*

Stofiel was named for Walter Stofiel, who first came to the area in the 1870s and became involved with mining at Island Mountain with Emmanuel Penrod, one of the original locators of the Comstock Lode. In 1877 Stofiel married Penrod's daughter, Lydia. In 1880 the Stofiel family moved to Bellevue, Idaho, only to return a couple of years later, when they established a small ranch and stage station near the future site of Wildhorse Reservoir. With the onset of the boom in Gold Creek, the Stofiel stop became an important shipping point for supplies heading to the camp. A post office opened at Stofiel on June 11, 1891, and did not close until July 15, 1897. Lydia Stofiel served as postmaster until she suddenly died of pleuritic pneumonia in January 1897. The Butler family ran the stage station until it had outlived its usefulness in the late teens.

The construction of the Wildhorse Dam in 1936 brought many workers to the area. The dam was completed in January 1937 and was dedicated on September 6. Besides providing irrigation water for ranchers and farmers, the reservoir instantly became a recreation spot. A couple of businesses opened nearby, and the Wildhorse post office operated from July 3, 1945, to February 28, 1948. Josephine Ford was postmaster. Today, Wildhorse Reservoir remains a mecca for vacationers and sportsmen. There is a store and a gas station nearby.

Telephone Mining District

DIRECTIONS: Located 7 miles north of Gold Creek.

The Telephone Mining District was located at the head of California Creek, north of Island Mountain. J. W. McDonald organized the district in the summer of 1885, after he discovered the Telephone mine in May. A. Van Uleck and Mortimer Smith worked the Review, another mine in the area. Van Uleck later became Tuscarora's district recorder and eventually Elko County recorder. The Hicks District bounded the Telephone Mining District on the north, and Allegheny Creek bounded it on the south. McDonald built some cabins at the mine and moved his family there, but his high hopes came to nothing, and the mine produced little before it was abandoned in 1890. Only small dumps and a collapsed shaft are left at the Telephone mine.

Tulasco

DIRECTIONS: Located 2½ miles north of the Welcome exit on I-80, 8 miles west of Wells.

Beginning in 1869 Tulasco was a sidetrack station on the Central Pacific Railroad, but it changed locations in 1909 when the Southern Pacific Railroad straightened its tracks between Deeth and Wells and built a section house and water tank. In 1912 Tulasco became the terminus for the railroad spur built to Metropolis, and a small depot, a saloon, and a restaurant served travelers. When the spur was abandoned in 1925, Tulasco became a siding again and housed only a section crew. In June 1936 section head

Depot foundations at Tulasco. (Photo by Shawn Hall)

George Johnson died when he stepped in front of a train. L. B. Fairchild, the gang foreman, had been walking with Johnson and was also injured in the incident. During the summer of 1946 the section gang experienced many problems. In two separate fights among gang members, two men, George Ruffin and Murphy Corley, died. Today, Tulasco serves both the Western Pacific and Southern Pacific Railroads as a signal station and siding, but all that is left of the town are the concrete foundations of the section house and water tower. Just to the east of the foundations is the old railroad bed for the Metropolis spur, where the saloon was located. It is marked by piles of the broken bottles.

Northeastern Elko County

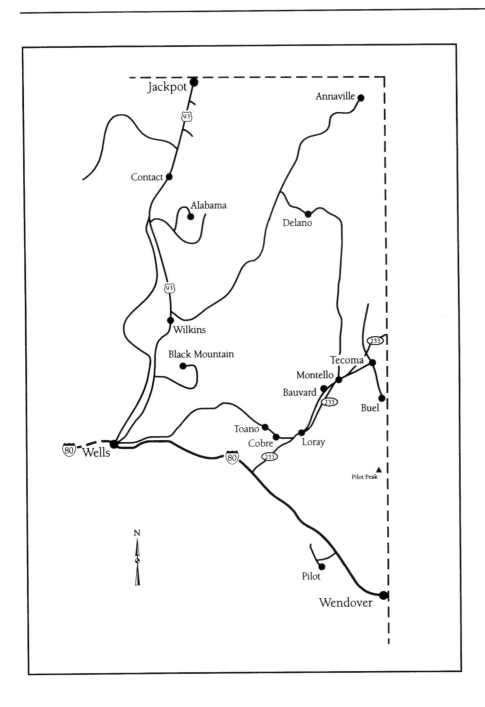

Alabama Mining District

DIRECTIONS: From Contact, head south on U.S. 93 for 8 miles. Exit east and follow the road for 10 miles to the Alabama Mining District.

Knoll and Slack organized the Alabama Mining District in 1871, after they discovered silver high on Blanchard Mountain. They called the discovery the Dayton and sent ore from the mine to Winnemucca for treatment. Knoll, an Alabama native, named the district after his home state. Water flooded the Dayton when it reached the fifty-foot level, and it was abandoned. During the next twenty years, a number of small mines were discovered in the area, including the Babel, the Arizona (also known as the Turo), the Blanchard (also known as the Bricker), the Alabama (also known as the Boston), the Johnson, the Vulcan, the Bell, the Silver Star, and the Silk Worm.

In 1876 prospectors discovered copper on nearby China Mountain. Eventually, mines in the Alabama Mining District were absorbed into the Contact, or Salmon River, Mining District. Alabama District mines produced consistently, but amounts produced cannot be determined because monetary returns were added into Contact production figures. However, the percentage the Alabama contributed to the total appears to have been small.

By the 1930s most of the area's mines had been abandoned, though production continued on a small scale until the early 1950s. The Sunshine Mining Company conducted some diamond drilling exploration in the region in 1972, and the Homestake Mining Company explored claims near the Vulcan mine in the mid-1970s, but no development took place.

Mine ruins abound in the district and are scattered within a six-square-mile area. They include head frames, collapsed buildings, and mine dumps. The remnants of a number of stone buildings are left at Babel, while newer buildings can be found around Blanchard Mountain.

Annaville

(Goose Creek Settlement) (Ashbrook Mining District)

DIRECTIONS: From Wilkins, head north on U.S. 93 for 1 mile. Exit right and follow the road for 46 miles to Annaville.

A small settlement with a trading post and lodging house developed at Goose Creek during the days of the Emigrant Trail, when Goose Creek was part of the Humboldt-Goose Creek-North Salt Lake branch of the California Trail. Goose Creek's simple amenities were a welcome sight for weary travelers who had just crossed the salt flats. The Chorpenning and Woodward mail stage ran through Goose Creek in 1853 and 1854, but the route was abandoned from 1855 to 1858 because of persistent Indian troubles.

The trail later reopened for emigrants, but Chorpenning and Woodward changed routes and ran farther south through Huntington Valley and Jacob's Well, a route the Pony Express later used. A number of emigrants decided to settle in the Goose Creek area. By the 1860s, when travel on the trail had come to an end, ranching was already an established way of life.

Some short-lived mining activity took place nearby, when the Ashbrook Mining District was organized in September 1871 after Alex White found gold high in the Goose Creek Mountains; the values were low and nothing was ever produced. A sufficiently large number of people lived in the area that a post office, named Annaville, opened on October 10, 1872, with James Hefferman as postmaster. Since Hefferman agreed to be postmaster, the residents did not object when he named the post office after his wife. The office closed on October 21, 1874.

Most ranches in Goose Creek failed in the twenty years following the area's establishment. By the turn of the century, only five or six were operating, and just two are still active. Old homesteads—symbols of man's early struggle to settle and earn a living in remote Nevada—are scattered throughout the Goose Creek area.

Bauvard
(Banvard) (Old Montello)

DIRECTIONS: *Located 3 miles southeast of Montello.*

The Southern Pacific Railroad established Bauvard, shown erroneously on some early maps as Banvard, in 1904. The town served as an engine terminal for trains traveling on the newly completed Lucin Cutoff. Residents quickly constructed a number of buildings to house railroad workers and shops for engine repairs. They also built a depot, a section house, and water tanks. Trains transported houses to Bauvard from Terrace, Utah, which had been isolated from the railroad by the Lucin Cutoff.

On June 4, 1904, a post office opened in Bauvard, with Wes Johnson as postmaster. However, within two years the budding town relocated to Montello, where the Southern Pacific had built a new division point complete with a seven-stall roundhouse. No buildings remained at Bauvard, and the town ceased to exist. Its name lived on for awhile, however, because though it was in Montello, the post office retained the name of Bauvard until February 27, 1912. Once the town had moved, locals referred to the site as Old Montello. Only sunken cellars and scattered lumber remain.

Black Mountain

DIRECTIONS: *Located 18 miles northeast of Wells.*

Black Mountain is a mining district that has seen little activity or production. The only known activity at Black Mountain took place in 1907, when James Cord worked the Dig and Pass claims and removed ore containing small amounts of lead, silver, and gold. After Cord left in 1908 Black Mountain was never disturbed again. Only a small collapsed shaft and mine dumps remain.

Buel

(Buell) (Lucin) (Tuttle)

DIRECTIONS: *From Tecoma, head south for 3 miles to Buel.*

Prospectors made initial discoveries in the Buel District during the summer of 1868. By 1869 the Yellow Jacket, the Wardell, the Central Pacific, and the First Extension mines, located on Copper Mountain just inside the Utah border, were producing copper. Silver discovered on Tecoma Hill, on the Nevada side, led to the formation of mines including the Tecoma, the Independence, and the Uncle Sam.

By 1870 a small town began to form, and a stage line to Tecoma, run by Billings and Ellis, started up. David Buel, who had achieved fame in Austin, arrived and bought most of the mines on Tecoma Hill for $125,000. During the summer of 1871 the town, now known as Buel, boomed. A townsite was laid out and within six weeks the town grew from five rough cabins to two hotels, three restaurants, six saloons, and twenty frame buildings. Businesses included the Joseph Andrews store, the Groepper Restaurant, the James Leffingwell Saloon, and a brewery built by Adolph Finck, which did a brisk business supplying the local saloons. A post office opened on December 18, 1871, in the Andrews store, with Andrews as postmaster. Three new stone quarries provided foundations for new buildings. New mines included the Revenue and the Irish American, both of which E. W. Markee, John Finley, and William Harrah owned.

David Buel was known in the West for developing a new style of smelter with a high extraction rate. In October 1872 the American Tecoma Mining Company asked him to build the Buel City Smelter. He completed and fired the $40,000, twenty-ton-per-day smelter on December 3 and sold it that summer to Howland and Aspinwall of New York, owners of some of the local mines. By the summer of 1872 a number of new mines began production; they were the Jeremiah Thompson, the Lucy Emma, the Helen, the Kentuck, the Osceola, the Molly, the Tempest, the Black Warrior, and the Growl.

Hoisting works of the
Mineral Mountain
Mining Company, 1908.
(Elizabeth Pruitt
collection, Northeastern
Nevada Museum)

Seventy men burnt charcoal to supply the smelter, and another 120 worked in the mines.

Buel peaked in 1874. During that year, forty tons of ore from the Black Warrior mine netted $16,000. Other mines producing included the Niantic, the Sharly, the Hattie, the Tecoma, and the Raymond. This was the last big spark for Buel during the 1870s, as one by one the mines closed. Businesses followed suit, many of the residents left, and the population shrank from almost 200 to less than 50 by 1876. However, in 1877 Buel was still large enough to have a justice of the peace, William McMillan, who also served as postmaster, and a constable, C. F. Rolland.

The Buel City Smelter closed in 1876, after processing several thousand tons of ore. The English Tecoma Mining Company, which now controlled most of the mines on Tecoma Hill, shipped 1,000 tons of ore to its smelter in Truckee, California, before leaving the district in 1878. This ended all mining activity at Buel for a time, and the post office closed for good on January 7, 1878.

In September 1885 a thirty-year-long revival began at Buel. It had become economically feasible to mine copper, now recognized as a valuable commodity. The first mines to produce were the Paymaster and the Shantie. During the next ten years, consistent production occurred. In 1894 the Salt Lake Copper Company bought a majority of the copper claims in the district. The company's confidence in its holdings prompted renewed interest, and during the next twenty years more than fifteen companies worked mines around Buel. By 1897 about fifty people lived in the Buel area, and a school

Mining buildings left at Buel. (Photo by Shawn Hall)

opened, with Isora Stevens of Starr Valley as teacher. The district was renamed the Lucin Mining District, but people still referred to it as Buel. While a number of mines were active by the year 1900, those in the Copper Mountain group, which the Lewisohn brothers owned, were the most productive.

The biggest boost to Buel occurred in 1907, when the volume of ore produced convinced the Union Pacific Railroad to build a spur from Tecoma, about four miles away. At the same time, the Buel Copper Mining Company built a four-mile tramway from the top of Copper Mountain to the railroad spur below the old Buel townsite. The railroad named the terminus Tuttle after a railroad representative, but people living there still knew it as Buel. The railroad completed the spur on September 6, at which time 150 men were working the mines. Many families moved to the town, and the school stayed filled to capacity.

The spur kept busy hauling ore, particularly when the mines on Tecoma Hill also began producing copper. From 1906 to 1912 they produced copper worth $1.7 million. By 1909 the Salt Lake Copper Company was producing 250 tons of ore a day from a double compartment shaft on Copper Mountain. Other mining companies working the district included the Little Butte, the Iron Mask, the Mineral Mountain, the Utah Lead, the Tecoma Consolidated, and the St. Lawrence. In January the old Tecoma Mines Company property, adjacent to the Salt Lake Copper Company mines on Copper Mountain, re-opened. The group also bought a controlling interest in the Iron Mask, the Mineral Mountain, and the Little Butte mining companies. In May Frank Williams reopened the Black Warrior mine and later sold it to the St. Lawrence Mining Company, after hitting a rich vein in October.

Between 1906 and 1917 a couple of businesses opened in the town, but most supplies were bought in from nearby Montello or Tecoma. Peak population during this period was about 200, but after 1917 most of the mines had closed, with only the Salt Lake Copper Company continuing to produce sizable amounts. The Buel school burned in 1923, when a clogged flue caught fire. An empty building in town was later used as a school but operated only occasionally until the school district was officially abolished in 1937.

Ore amounts continued to fall, and the Salt Lake Company folded in 1935. The district had grown silent, with only empty ore buckets hanging from the tram line providing testimony to Buel's past glories. The railroad spur, unused for a number of years, was officially abandoned on May 1, 1940, and the aerial tram was sold for scrap in 1941. The last major bout of production took place in 1953 when MacFarland and Hollinger mined $50,000 worth of low-grade limonite for the Atomic Energy Commission. While no production has occurred in Buel since, there has been extensive exploration in the area. In the 1980s the Union Oil Company drilled numerous test holes, and the Moly Corporation tested Sixshooter Canyon from 1980 to 1986. Currently, drill rigs are again testing the district, looking for disseminated gold deposits. Total proceeds of production from Copper Mountain and Tecoma Hill stand at around $5 million and include more than twenty million pounds of copper.

There is still much to see at Buel. The Tuttle townsite, where a couple of original buildings remain, is located on a private ranch. A huge pile of wood planks is all that is left of the tram house, while a cemetery is located a quarter of a mile away. On Tecoma Hill, ore bins, mine shafts, foundations of the Buel City Smelter, and numerous buildings are scattered along the winding road. A few collapsing log cabins mark the old Buel townsite. The highlight of Buel is the tram. While the ore buckets are gone and wires are half buried in the sand, almost all the supports remain. The road to the Copper Mountain mines is rough and steep, but definitely worth the effort necessary to navigate it. From the top of the mountain, the tram line is visible, straight as an arrow, right to Tuttle. On the other side of Copper Mountain is a spectacular view of the Utah salt flats. Buel is a definite must-see. Plan to spend a full day there, and bring your camera.

Cobre

(Omar)

DIRECTIONS: From Oasis, take Nevada 233 north for 7 miles. Exit left and follow the road for 1 mile to Cobre.

Cobre, originally named Omar, served as a station on the Elko division of the Southern Pacific Railroad. The actual Omar station was located some distance to the east, but when Cobre was chosen as the terminus for the Nevada Northern Railway, most of the town's buildings were moved to the new townsite at Cobre.

Omar achieved no prominence until work began on the Nevada Northern Railway in 1905. Because of the link with rich copper mines in the Robinson district in White Pine County, the railway terminus was named Cobre, Spanish for copper. W.H. Wattis supervised grading for the railroad, and it was completed to Cobre in September. Actual railroad construction from Cobre began in November.

Cobre boomed in 1906 because the Utah Construction Company laid track, and all supplies came through the town. The Western Pacific Railroad moved its construction headquarters from Winnemucca to Cobre. In February Cordelia Spencer opened the Pioneer Hotel, and a post office opened on March 12, with Spencer as postmaster. In 1907 Spencer sold her hotel and other holdings in Cobre to Horace Kelley for $25,000.

As Cobre grew, it developed a reputation for violence. In June 1906 an insane man, R.A. Hanlon, took three shots at J.F. McBride, a saloon owner. Harlon missed, but McBride did not. McBride's three shots found their mark, killing Hanlon. The death was ruled justifiable homicide. Another incident in August 1907 also involved a local drinking establishment. Two masked men robbed the J.C. Hillman Saloon of $1,200. The saloonkeeper, G.B. Gilliam, drew his gun, but the robbers shot him through the head and he died. A sheriff's posse employed local Indians to track down the men. They killed one man, but it turned out that he was not one of the murderers. Locals complained that the sheriff was incompetent and had the posse proceed unaided, but the killers were never caught. Investigators believed that Gilliam recognized his killer as an ex-con from when he had been a turnkey at the Washington State Prison. Gilliam and Hillman had received three letters warning them that their lives were in danger.

In 1908 Hillman sold his saloon to Horace Kelley, who now owned most of the businesses in Cobre. Hillman decided to concentrate on his mining interests in Six-Mile Canyon, ten miles west of Cobre, where he joined J.A. Smith and organized the Cobre Lead Mining Company. The company constructed a number of buildings at the mine, but mining activity lasted only a short time.

Passenger and freight depots at Cobre, 1930s. The Southern Pacific tracks are in the forefront, Nevada Northern tracks are by the water tower. (Mel and Mae Steninger collection, Northeastern Nevada Museum)

By 1910 Cobre boasted a population of sixty. Its businesses included a hotel, a store, a mercantile company, and three saloons. Cobre had reached its peak. While ore trains from Ely kept coming through Cobre, passenger and freight traffic gradually declined during ensuing years. However, the people that remained necessitated the building of a school in 1915. The first teacher was Amy Parker. The school remained open through the 1930s. In December 1921 the town was shocked by postmaster George Hall's arrest for embezzlement and issuing money orders with no funds in the office. He had lost money in a local poker game and wrote money orders so that he could continue playing. Mayme Mitchell soon replaced him. The post office closed on May 31, 1927, but reopened on June 10, 1929, with John Toyn, owner of the Cobre Hotel, as postmaster.

By 1937 Cobre was being labeled a ghost town even though twenty people still lived there. In 1930 a large pumice deposit a mile and a half north of Cobre began producing, but it was not until the summer of 1936 that the Cobre Minerals Corporation was organized. Extensive development began, and the company made plans to build a large mill. The company, based in Detroit, claimed it had the only deposit of true pumice in the United States. The *Wells Progress* declared that, "Ghost Town of Cobre to become Important Manufacturing Center of State."[1] The Bowers Building and Construction Company of Salt Lake City began construction of a $500,000 mill in late 1937. The company constructed about a dozen buildings near the mill, and the pumice mined was used to make a variety of products including acoustical plaster, synthetic travertine, marble, tile shingles, building blocks, and fire bricks. The 218- by 60-foot mill was completed in May 1938. Its daily capacity was 150 tons. The mill employed 15 men, although the number rose to nearly 100 by the end of summer. The company held to a strict policy of hiring only men from Nevada. The housing built for employees had baths, showers, toilets, and electric lights. But markets for pumice were weak, and the operation shut down after a few years with only a minimal profit to show for all the effort.

Cobre returned to its past ghost town status, with only a handful of people still residing in the town. The widespread availability of automobiles and trucks made the Nevada Northern Railway obsolete, except for hauling copper ore. Daily passenger service ended to Ely in 1938, and all service came to a complete stop in July 1941. In November 1948 the Southern Pacific Railroad abandoned the Cobre station, and the Cobre Mercantile Store took over the handling of any railroad business. Cobre's end arrived on May 31, 1956, when the post office closed for good.

The Nevada Northern Railway used Cobre as a shipping point until the 1980s, but the trains rumbled through a dead town. As recently as 1980 a couple of original buildings remained at Cobre, but only foundations are there today. The only structure left is a cinder block engine house built during the last years of the Nevada Northern Railway. At the pumice mine, extensive mill foundations and other ruins mark the site.

Contact

(Contact City) (Salmon City) (Salmon River)
(Kit Carson) (Porter) (Portis) (Alabama)

DIRECTIONS: *Located 16.4 miles south of Jackpot on U.S. 93.*

While Contact did not become prominent until after the turn of the century, the first discoveries there took place in 1870, when James Moran found gold. No production occurred, however. In 1872 a prospecting party of three men (Hanks, Lews, and Noll) located a number of new deposits on China Mountain. As a result, other mines opened. By the end of that year they included the Pocohantas, the Golconda, the Dunderberg, and the Virgin. Four separate but adjacent mining districts—the Salmon, the Kit Carson, the Porter, and the Alabama—were organized. Mines in the Kit Carson District were the Juniper, the Montezuma, the Edith, the Morningstar, the Polar Star,

and the Sioux. Servis also served as district recorder. All the mining districts were later consolidated into the Contact Mining District. Although a town did not form, a small hotel was completed at Contact in April 1874 and served as a stop on the Toano and Idaho Fast Freight Line.

In 1876 an official of the Southern Pacific Railroad located a number of other deposits on China Mountain, and a large number of Chinese worked the mines on a commission basis. The most prominent of the early mines, the Boston, shipped copper ore to Swansea, Wales, until 1880. During the early 1880s new mines continued to open, but sustained production was minimal. In the Kit Carson District, Thomas Cochran and John Mitchell opened the Reverend Goode and the Exchequer Goode, and L. F. Wrinkle worked the Cedar, the Nevada, and the Stormy mines. In the Porter District, W. J. Hanks and P. H. Jackson discovered the Arizona mine and relocated the Boston, Albany, and Harper mines. The Porter, the Utah, the Dividend, the California, and the Juniper mines were also active.

The Stevens Gold and Copper Mining Company owned the Ontario, the Michigan, the Erie, and the Elko mines. In the Alabama District, miners worked the Lizzie and the Emily. The Salmon River District mines were the Domingo, the Blue Jay, the Manhattan, the Riverside, the Chester, and the Wonder. Despite this flurry of activity, the mines produced little. By the end of 1883 most people had given up, and the area was essentially abandoned until 1887.

In 1888 Hickey, Delano, Ayres, and Hechathin located the Delano group, which eventually became one of the most productive in the Contact District. In 1889 Warwick and English located the Brooklyn mine, and by 1891 two other mines (the Empire and the Copper Queen) had also begun production.

By 1895 Contact mines employed seventy miners, and the extensive prospecting around the town continued to yield new discoveries. In 1896 David Bourne, later known as the father of Jarbidge, located the Mammoth mine, and Coleman, Moore, and Thompson discovered the Bonanza. Coleman later became rich after discovering the Father De Sinet property in the Black Hills.

Eighteen ninety-seven was a year of great promise for the town. In January 100 men were working at Contact, and by April the number had risen to 200. A post office, with Eugene Shields as postmaster, opened on February 6 of that year. A number of mines, including the Blue Bird, the Jackrabbit, the Reliance, the Yellow Girl, and the Delano were being worked. Twenty miners who contributed $100 each for the purpose of building a fifty-ton smelter formed The Salmon River Mining Company. But the smelter proved extremely unsuccessful and only processed fourteen tons of ore from the Bluebird mine before closing. This was a heavy blow, and many people left the area soon afterwards. In February 1898 the Delano property was bonded to an English syndicate for $65,000 but nothing came of it. By 1900 Contact's population had dropped to eighty-five, and only five residents were left by 1905.

A slow revival beginning in late 1905 led to Contact's first real boom. A new town began to form just south of the original townsite, which had contained only a couple of buildings, including a school, with Flora Vincent as teacher. A stage line to Twin Falls started service in September 1906. The new town was named Contact, a mining term for the contact zone between the granite and porphyry in the district. The United States Mining and Smelting Company came to the district in 1907. By 1908 the population stood at 300. In February the Contact Power and Milling Company, based in Seattle, bought many claims and immediately made plans to build a $200,000 concentrating plant with its own power plant. Prospects looked so good that in April 1909 three townsites: Contact, Contact City, and East Contact, were platted. Two companies—the Western Townsite Company (Mose Jones) and the Contact City Townsite Company (Henry Smith)—developed the townsites, and Smith declared Contact the "Butte City of Nevada."[2]

The most developed of the towns was Contact City, which contained fifteen buildings. Between the three townsites, 450 lots were sold by 1915. The thirty-five-room Contact Hotel, built with local granite, was completed in August. A number of saloons opened, including G. L. Collins's Mint Saloon and Restaurant, the Palace Saloon and the Northern Saloon, and the Blue Ribbon Bar. A barber shop, and the W. A. Kent General Merchandise Store also opened. Kent was called the "Pioneer Merchant of Contact." Residents loved him.

A promotional paper, the *Contact News*, began publication on May 20, but its nature is a mystery. Historical books do not mention the paper, and the fact that it existed was not known until a copy of the third issue was given to the Northeastern Nevada Museum. This issue reveals little about the paper except that it appears to have been printed by the Western Townsite Company. The publication was short-lived and apparently did not survive the summer.

In 1910 the Nevada Copper Mining, Milling, and Power Company, based in Tecoma, bought out the United States Smelting Company. The Contact Company was renamed the Contact-Seattle Copper Mining Company, and the two organizations controlled virtually all the mines in the district. Because of the demand, a tri-weekly stage to Rogerson, Idaho, began in April. Another newspaper, the *Contact Miner,* began publication on March 20, 1913. J. V. Marshall served as editor and owner, under the auspices of the Miner Publishing Company. The staunchly democratic newspaper cost $2.50 per year. In December 1915 Marshall sold out to E. H. Childs, but Childs only printed three or four issues of the paper before ceasing publication.

In August 1914 the Nevada Copper Company began construction of a 100-ton leaching plant and named Henry Smith manager. Coincidentally, Smith had sold the land for the plant, which the company completed in late 1915. The company spent most of that year developing four mines (the Delano, the Champ Clark, the High Ore, and the Copper Shield), but they produced only $2,300 worth of ore. In 1916 the Contact-Seattle Company's Delano mine produced almost all the $181,000 worth of ore mined in the area. The mine's ore kept the mill running and the competition solvent. Contact mines produced ore from 1916 to 1958, a tremendous record of consistency. In 1918 the Vivian Tunnel Company began work on a new mine, a Sutro-type tunnel intended to help drain water from the Contact mines and provide an easier method of ore removal. However, the project proved very expensive and was abandoned before its completion.

While consistent production occurred from 1916 to 1919, the 1920s was the decade that put Contact on the map. In September 1922 the Three in One Mining Company announced plans to build a $2.5 million smelter fifteen miles south of Contact. However, it was revealed that the plan was an investor scam, and the smelter was never built. Despite this, Contact continued to prosper. In 1923 H.A. DeVaux, head of the Contact Sewerage Company, constructed a number of new buildings, including a two-story, thirty-room office building built with Contact granite. By 1924 the town was buzzing over construction beginning on the Union Pacific Railroad's Oregon Shortline, which ran from Rogerson, Idaho, to Wells. A jail opened in March, and overindulgent miners filled it to capacity on the weekends.

With great fanfare, the $110,000, three-story, fifty-room Fairview Hotel opened on May 24, 1924. One of the owners, H.A. DeVaux, provided fireworks, and W.G. Greathouse, Nevada secretary of state, was guest of honor. The Contact Construction and Investment Company, of which Robert Weir was president, owned the hotel. In July the Gray Mining Company, of which the Vivian Tunnel Company was a subsidiary, began to work the Vivian Tunnel once again, hoping to quickly complete the 20,000-foot tunnel. Mark Musgrove, former mining editor of the *Nevada State Journal,* began running a new newspaper, the *Nevada-Contact Mining Review,* on November 22. The *Nevada State Herald* published the paper, but Musgrove never paid the *Herald* for the printing. *Herald* editorials, which had originally praised the Contact paper, began to sullenly assert that the paper was a blight on the newspaper business. Musgrove switched the printing job to a plant in Filer, Idaho, in January 1925, but in June Leslie Fox, who also ran the Kimberly (Idaho) *Tribune,* took over the paper. Despite the new ownership, the paper folded in September, to the great glee of the slighted *Herald.*

On March 11, 1925, the Oregon Shortline's first construction train arrived in Contact, and a depot, located below town next to the Salmon Falls River, was built. Regular service began within a month. Meanwhile, the Fairview Hotel had fallen on hard times and went into receivership in September. Robert Weir Jr. initially bought it at a tax sale for $1,525, but the ploy did not work, and the owner's son's bid was disallowed. At another sale in October, J.L. Newland purchased the hotel for $2,700. The hotel caught fire on May 31, 1926, and it took only thirty minutes to completely destroy the expensive building. By the end of 1926 the main business left in Contact was the Fred Johnson Merchandise Store. Johnson also owned branch stores in Wells and Montello. A new school opened in 1927, and the two-room building housed forty students. Flo Reed, who taught at many Elko County schools, served as teacher from 1927 to 1930. Marguerite Patterson Evans was the school's last principal and the only teacher in the high school, which closed in 1934, though it saw sporadic use during Contact's revival periods. In the 1920s and 1930s two separate schools operated. The grammar school was the two-room building constructed in 1927, while the high school occupied an old pool hall.

The advent of prohibition in 1917 did little to dull Contact residents' thirst. Bootlegging was widespread in the mountains of the area, and rumor had it that in the 1920s there were more bootleggers than there were people living in the town. The biggest bootlegging operation was at a place called Heaven's Delight. Hell's Delight, a speakeasy in town, sold Heaven's Delight's products. Virgil Church recalled that "there were six major moonshine operations in Contact from 1917 until the repeal of Prohibition in 1932. Those were just the big outfits. I was a moonshiner. Hell, everybody in town was a moonshiner making grain whiskey in their cellars."[3] John Detweiler echoed Church's sentiments.

> My father was justice of the peace here for years. He was never a bootlegger or moonshiner. Said he could not risk the chance, said they'd throw away the keys to the jailhouse if the judge was ever arrested. But hell, he was the No. 1 supplier of everything needed to make the whiskey. Dad hauled in all the coal, barrels, wheat, and sugar to supply the moonshiners. Dad had the slot machine concession in town and would always take me with him when he collected the money from the slots at Hard Rock Tilly's sporting house at the south end of town. As long as I was with him, mother figured dad would not get in any trouble.[4]

In 1930 a new Contact townsite was laid out. A power plant that the Vivian Tunnel Company had built provided power for the town. Contact's population stood at 260, but the town's heyday was over. In October 1931 the Gray Mining Company sold its property and buildings for taxes. An absence of bids forced Elko County to buy the property for $956, though its appraised value stood at $37,000. In January 1932 the property sold again for taxes, but Elko County had to buy it once more, this time for $1,124.

In 1935 Contact still had two general stores, a hotel, two saloons, a post office, and a school, but the population had shrunk to 100. The WPA program supported most of the people remaining in town. A depressed copper market plagued Contact, and mine production continued to falter. Despite the slowdown, an Episcopal church, St. Agnes Chapel, was completed and dedicated in June 1936. In August 1942 the church fell victim to a devastating fire that started in the L. C. Bugbee Mercantile Store, when a kerosene burner in a refrigerator exploded. There was hardly any water to fight the fire, and the building, which housed a store, a restaurant, a hotel, a bar, and a service station, was completely consumed. The flames spread to the church and also burned a number of homes before being extinguished.

Mines around Contact did not revive until 1943, when war demand for copper raised prices and made mining profitable again. The Marshall Mining Company started working the Delano mine and developing a new prospect, the Marshall mine. At the same time, W. C. Lewis and Charles Whitcomb began development of the Bonanza property. Fire visited Contact again in May 1947, when a service station, a bar, a store, and a dance hall, all of which Ed Henzinger owned, were completely destroyed at a cost of $40,000. In November Contact elected Ray King as its first mayor. Despite renewed optimism, the Contact revival had faded completely by 1947. From 1943 to 1946 Contact's mines produced 800,000 pounds of copper, but from 1947 to 1951 they yielded only 7,000 pounds. Another fire in August 1951 destroyed the last hotel, formerly owned by Contact mayor Ray King.

Contact's last revival began in 1952, when the Marshall mine was renamed the Nevada-Bellvue. Extensive work ensued. But this new activity came too late to save the Contact stop on the Oregon Shortline, and the Union Pacific closed the depot in December. The depot later moved to Lee, south of Lamoille, where it still stands today.

A strange occurrence took place in Contact in March 1953 when seventy-year old Thomas Williams shot and killed seventy-year old J. R. "Tex" Hazel-wood. The two had been feuding for years. Since the 1920s Tex had been building himself quite a reputation around Contact and nearby ranches of the Union Cattle Company, where he was labeled as being "one stave short of being round." He roamed the area living in caves or crude willow shelters, and his reputation for strangeness kept him off all local ranch payrolls. He achieved the height of his notoriety when he came up with a new way of using shoes with cow hooves on the bottoms. He made the shoes himself. Puzzled cattlemen could not figure out how their cattle were disappearing, though there were no human footprints nearby, even when the snow was deep. Lawmen finally caught Tex, and he spent a couple of years in prison. Once he was released, he returned to the Contact area, where he continued to cause problems. It was because of his orneriness and his shenanigans that a disgruntled acquaintance eventually killed Tex while he was sitting in his pickup truck in Contact. His rustling shoes are on permanent display at the Northeastern Nevada Museum.

The Nevada-Bellvue mine, still owned by Maurice Marshall but leased to the American West Exploration Company, produced substantial amounts of copper from 1952 to 1957. Ore, which continued to be shipped from the Contact siding despite the depot closure, went to the huge smelters at Garfield, Utah. From 1952 to 1957 the Nevada-Bellvue produced virtually all the two million pounds of copper it mined. It has yielded nothing significant since. All told, the Contact District produced 5.8 million pounds of copper, 360,000 pounds of lead, 127,000 ounces of silver, 18,000 pounds of zinc, and 1,200 ounces of gold. By the time the Contact post office closed on August 31, 1962, the town was virtually empty.

In the 1970s the Sunshine Mining Company explored the Contact area, as did Exxon Minerals and the Homestake Mining Company in the 1980s, but the area has seen no other activity since the fifties. A few residents still living in Contact work mainly in Jackpot or for the highway department at the Contact Maintenance Station. In the oldest part of Contact, located just north of the maintenance station, are the rock walls of one of the town's first stores. In the main town of Contact, west of U.S. 93, old homes stand alongside newer trailers. An impressive concrete building, which served as the community social hall, dominates the townsite. Many dances and other social functions took place there, and the walls were covered with grandiose murals depicting the developers' fanciful visions of the Contact of the future. The old school, now a residence, remains. A cemetery is located nearby, and many reminders of Contact's mining heyday abound in the surrounding hills. The rails of the Oregon Shortline were torn up after the line folded in 1978.

Delano

(Delno) (Goose Creek Mining District)

(New York Mining District)

DIRECTIONS: *From Montello, head north for 14 miles. Take the right fork and follow it for 10 miles. Turn left and follow the road for 8 miles. Turn left again and follow that road for 1 mile to Delano.*

Thomas and Brown discovered mines at Delano, or Delno as it is more traditionally known, when an Indian led them to a deposit there in 1872. The Goose Creek Mining District was organized in July, and ninety claims became active. In May 1875 the district was renamed the Delano Mining District, in honor of Alonzo Delano, an ancestor of Franklin Delano Roosevelt. Delano wrote the *Overland Journal of 1849*. Delano's active mines during the early 1870s included the Eliza, the Southern, the Slavonia, the 74, the Arch, the Elko, and the Eastern Nevada. Johnston also served as district recorder. The Chicago Gold and Silver Mining Company gained control of many mines in 1872. The Servia and Slavonia Mining Company's incorporation in 1876, with directors William Holden, H. P. Irving, Charles Abbott, and Samuel Morrison, raised great hopes. Abbott also served as the Delano justice of the peace. With a capital stock of $10 million, it appeared that Delano would soon be mined extensively, but that was not the case. The company quietly disappeared a couple of years later without mining the district.

Limited mining continued during the 1870s and 1880s. Mines that opened during this period included the Don Pedro and the Bay State, the New York, the Louisa, the Flagstaff, the North Star, the Sophea, the Fourth of July, and the Holy Terror. In 1884 twelve miners were working a number of mines. The main producer was the Garfield, which Argyle and Company owned and which shipped the silver-lead ore to Salt Lake City. The New York Mining

District, adjoining the Delano district to the south, was also organized. Its main mine was the Silver Belt. D.W. Ranch, a Tecoma deputy sheriff, had a one-third interest in the mine. Enough interest was exhibited in the district that in 1888 Dake planned to build a mill at Annaville Spring, two and a half miles south of Delano, but the water source proved inadequate, and plans were scrapped.

While a camp had formed over the years, a saloon was the only business in Delano, and supplies came in from Tecoma. By the summer of 1891 Delano's population was around thirty. The main obstacle hindering Delano's development was a lack of water, since the nearest source was more than three miles away. A small boom began in 1892 when Craig Chambers hit a rich vein in the Gold Note mine, which Dake and Johnston discovered in 1884. In the next ten years the mine produced close to $100,000 worth of lead and silver, and David Stonebreaker built a twenty-ton concentrator on Rock Spring Creek, located seven miles west of Delano. However, by 1903 the mines were idle, and only limited claim work took place until new deposits in the Cleveland mine began production in 1908. Joe Moran, N.M. Pratt, and J.N. Johnson had originally discovered the mine in 1886. By the end of 1908 the Cleveland and Argyle mines remained active, but production was erratic until 1916. From 1916 to 1970 the mines produced a combination of silver, lead, copper, and zinc. By 1918 three mines (the Net, the Gold Note, and the Cleveland) yielded ore valued at $38,000, and a year later seven mines generated revenue in the amount of $89,000. The district's best producer was the Net mine, which the Delno Mining Company owned. In 1921 the Panther

Gold Note Mine, Delano. The ore loading chute is from the early days of Delano. This, and most other remains, burned in August 1996. (Photo by Shawn Hall)

Building in the main section of Delano, destroyed by fire in August 1996. (Photo by Shawn Hall)

Mines Company discovered the Delano mine, which was destined to be one of the district's biggest producers. Ore had to be hauled by wagon to Granite Creek and then loaded into an ore bin for transfer to trucks. The company had Wes Johnson build a twenty-ton concentrating mill in 1926, and the following year had it expanded to fifty tons. The mill was adjacent to the old mill on Rock Spring Creek.

The population of the camp had grown to about fifty, and a number of buildings were crowded onto the flats below the Cleveland mine. As a result, on May 5, 1927, the government granted the town a post office named Delno. Cloe Richards was to be postmaster. However, the government rescinded their offer to supply the town with a post office before the office opened, because by time the forms, filed a year before, had gone through government channels, Delano was experiencing a lull. But the district was back on track by 1929 and 1930, when mines produced more than $114,000 worth of silver, lead, copper, and zinc. In late 1930 lack of water forced the closure of the mill at Rock Springs for the first time, and low metal prices led to the mines' closure in April 1931. Little was mined during the next three years.

It was not until 1934 that two mining companies acquired virtually all the mines of the district. The Delno Mining and Milling Company and the United Metals Company worked the Cleveland, the Net, and the Panther mines. In August the fifty-ton mill at Rock Springs was dismantled, because it proved cheaper to ship ore to other mills via Tecoma. A school opened in the fall and had five students. Tragedy struck in May 1936 when a diesel engine at the power plant exploded, and Ernest Canady of Montello was instantly killed.

Another near disaster occurred in January 1937 when two men, Jess Baker and Harris Collins, were trapped at the 300-foot level of the Cleveland mine. Luckily, long-time Delano resident Joe Mitchell remembered an old mine entrance that led to the trapped men, making their rescue possible.

Ore production was consistent through the mid-1940s, and the Delano District enjoyed its best years from 1946 to 1950. In 1946 the district's mines produced one million pounds of lead and 76,000 ounces of silver, valued at $177,000, and the Cleveland mine ranked fourth in Nevada for silver production. The mines generated revenue in the amounts of $157,000 in 1947, $341,000 in 1948, $207,170 in 1949, and $141,000 in 1950. While there were a number of good production years in the 1950s, nothing in that decade approached the production levels of the late 1940s. But even as late as 1960 the Delno and Gold Note mines were Elko County's largest producers of lead.

From the early 1920s to the early 1960s Delano's population stayed between thirty and fifty. In the 1960s production slowed considerably, and only about ten people lived in the town. From 1918 to 1970 mines in the Delano District produced 22 million pounds of lead, 1.5 million ounces of silver, 1.4 pounds of zinc, and 167,000 pounds of copper. While the last substantial bout of production took place in 1970, the district is still being actively explored. Noranda Exploration drilled the Delano and Cleveland mines, and Richfield Resources bought the Gold Note mine, which shows signs of recent work, in 1989.

Until recently there was much to see at Delano. Many buildings of various ages were scattered throughout the area. The main group, just above the Delano mine, included the school. Other buildings were left just below the Cleveland and Gold Note mines. In addition to more than twenty buildings remaining at the site, a number of headframes still stood at the mines. The Delano mine boasted an incline shaft complete with an engine house with all its equipment intact. However, a 40,000-acre fire struck the ghost town in August 1996, destroying all but a couple of buildings. One of Elko County's best ghost towns is now gone forever.

Loray

(Luray) (Leroy)

DIRECTIONS: *From Cobre, head north on Nevada 233 for 6 miles to Loray. The mines of the district are located 2 miles southeast of the Loray siding.*

Loray was first a station on the Central Pacific Railroad and later served the Southern Pacific Railroad, but the station had few passengers and was used primarily to ship wood cut nearby for railroad use. Wood crews lived in housing next to the tracks. A near disaster took place close to the station in May 1875, when two freight trains collided. The collision wrecked three engines and threw twenty-five cars off the tracks. Brakeman William Cassin and conductor Snyder were seriously injured but eventually recovered. In 1883 A.P. Shively, also the district recorder, organized the Loray Mining District as a result of some copper ore discovered about three miles south of the railroad in the early 1880s. During the next few years miners worked the Shively and Wilson, the Castle Park, the Stokes, the Queen of Sheba, the Birthday, the Ruby Chief, and the King Fisher, though they produced little until after the turn of the century.

In March 1907 twenty tons of ore from the Anderson property brought back promising returns. R.P. Christian was in charge of eight men and the operations. A copper and silver strike occurred on the Will Porter claim in May 1908, and Kellough and Lang discovered the Lillian mine in July 1908, but the district's mines produced only intermittently throughout the district's active years that extended to 1958. Mines that produced between 1906 and 1958 included the Alabama, the Jay Bird, the Pickup, the Maybelle, the Castle Peak, the Lost Hope, the Silver Bell, the Silver Star, and the New Deal. The Delno Mining and Milling Company worked a few of the mines sporadically from 1928 to 1936, when Charles Roberts reworked old tailings and recovered more than $69,000, primarily from the Maybelle mine, which Walter Long discovered in 1934.

During the 1930s and 1940s most miners lived at the Loray siding, where a school operated for a number of years. No production has been recorded in the district since 1958. Altogether, the Loray District produced 480,018 pounds of lead, 28,802 ounces of silver, and 25,883 pounds of copper for a total value of $191,000. Mining remnants abound on both sides of the mountain east of Loray. They include gallows frames, shafts, and even a converted Model-T that was used as a hoist. At the Loray siding, where a dozen buildings once stood, only foundations and scattered debris are left.

Montello

(Bauvard)

DIRECTIONS: *From Oasis on I-80 head north on Nevada 233 for 23 miles to Montello.*

The Southern Pacific Railroad created present-day Montello in 1904, when it built the Lucin cutoff across the salt flats of Utah. But long before that, a train robbery—a rare occurrence in Elko County and Nevada history—took place in the area. The robbery occurred in January 1883 when the Central Pacific Railroad was in operation. The engineer stopped his train when he saw a red flashing light on the water tank. Masked bandits immediately set upon him and his crew, tied them up, and put them in the siding

Montello railyards, 1914.
(Edna Patterson
collection, Northeastern
Nevada Museum)

shack. The bandits headed to the Wells Fargo express car, where messenger Aaron Ross was protecting the shipment. Ross steadfastly refused orders to come out and barricaded himself in the car where, despite being hit by three bullets, he continued to defy the robbers. Another train approached the siding, and the robbers met it and ordered the crew to continue on. With guns at their heads, the crew could do little. The robbers returned to work on the express car, and when they tried to force their way in again, Ross shot one of them. They then tried to smash the car by ramming it with the engine, but fears of another train coming forced them to abandoned the attempt. The only booty taken was ten dollars they found in the conductor's wallet. Sheriff Henry Taber of Elko organized a posse. With a reward of $1,250 for each man, this was an easy task. The first two robbers, Orris Nay and Frank Hawley, were wounded and captured in Utah County after a shootout on January 28, and the pair told the posse where the rest of the gang was hiding out. The other three surrendered without incident when the posse, which was armed with dynamite and a small cannon, confronted them. The posse took Sylvester Earl, Erastus Anderson, and Frank Francis, along with the other two robbers, to Salt Lake City. The prisoners were returned to Elko in February, and all pled guilty. Nay, Hawley, and Francis were sentenced to fourteen years in prison, while Earl and Anderson, who were both teenagers, received twelve-year sentences. Ross was celebrated as a hero for refusing to surrender to the robbers and received rewards from the Central Pacific Railroad and Wells Fargo. The rewards for capture were divided among the seventeen men of the posse, each of whom received $350.

The first train came through Montello on March 4, 1904, after the Southern Pacific Railroad created the town, and houses were moved to Montello from Kelton and Terrace, Utah. Montello replaced Terrace as the main railroad division point, forcing the town of Terrace into oblivion. Montello was initially known as Bauvard, also the name of the old siding located three miles to the southwest. The old Bauvard post office was transferred to the new town, but the name was not changed to Montello, which is Shoshone for rest, until February 27, 1912. The railroad built a large hotel, which burned in 1908, as well as complete railroad facilities at Montello. A small Chinatown developed because many of the Chinese roadbed and maintenance crews were based at Montello, and the forty residents supported a store and joss house. In July 1904 O.T. Hill finished construction on an eight-room house, storeroom, and lodging house for Wes Johnson, who also opened a general merchandise store. In October a part of the new library was used as a school, and Ethelyn Allen taught ten students the first year.

A schoolhouse opened in 1906. In 1907 the Utah Construction Company purchased the Vineyard Land and Livestock Company's vast holdings. Because of Montello's convenience to the railroad and their ranches, the company made the town its headquarters. The company ran a store in Montello for many years, and the original store, displaying the UC logo, still stands there. From 1904 to 1929 Montello's population was close to 800. The town also served as a supply point for the mining camp of Delano, located to the north. The railroad payroll reached as high as $1 million a month in 1915. A well-appointed new school opened in 1910. The old school went downtown

Downtown Montello, 1995. (Photo by Shawn Hall)

and converted into the town's amusement hall, which the Montello Amusement Corporation ran. The Montello Mercantile Company was incorporated in July 1912. A.C. McGinty built a new jail in the spring of 1916, replacing the original railroad tie jail, from which anyone could escape by removing a tie or two.

Montello began to decline in the 1920s. A major blow struck the town when a large fire, started by an exploding coffee urn in the Nelson Cafe, burned the business district in October 1925. It destroyed the Wes Johnson store, the Nelson Cafe, the Bryant Pool Hall, the post office, and the home of A.R. Cave. In addition, the Utah Construction Company split up and was sold. A portion of the holdings, the Gamble Ranch, is still one of the largest ranches in Elko County. When sold in 1950, the purchase price was $3 million. Montello received national attention in November 1928, when future president Herbert Hoover gave a campaign speech at the town. The following year the American Chiropractors Association elected local doctor J.D. Sherrod vice-president, and a KSL radio broadcast featured the Montello Consolidated School Orchestra in March 1936.

The introduction of diesel engines in the 1950s spelled the end of Montello's importance to the railroad. Soon afterwards, the roundhouses and shops were removed, and only the huge water tower was left behind. In 1960 a paved highway to Montello was completed from Oasis. Today Montello is a quiet town with a population of about seventy-five. Many of the homes moved from Utah in 1904 still remain, and the original depot has been relocated and converted into a home. The school, the jail, and other old business buildings are still there, and the cemetery is just north of the main street. The water tower dominates the old railroad yards, and trains continue to rumble through Montello. A store, a gas station, a post office, and a few saloons still operate.

Pilot

(Pilot Peak) (Pilot Peak Mining District)

DIRECTIONS: *Pilot Peak is located 18 miles north of Wendover.*
Pilot is located 10 miles northwest of Wendover.

Pilot Peak served as a landmark for emigrants crossing the salt flats of Utah on their way to California. John C. Frémont named the peak in 1845, a few years after the first emigrant group, the Bartleson-Bidwell party, came through in 1841.

The Pilot Peak mine, which F.H. Darling owned, was the first mine on Pilot Peak. In July 1878 Walter Brown made some small silver discoveries and organized the Pilot Peak Mining District. His mines were the Thistle and the American Flag. Brown was overly confident and also filed for a mill site, but not much activity ensued at his mines, and they produced little, if anything, in the 1870s. It was not until 1908 that interest in mining the district renewed. At this time J.C. Hillman, a saloon owner in Cobre, and William Porter began working some claims, but again, they had little to show for their efforts.

With the completion of the Western Pacific Railroad, a stop and signal station named Pilot came into being. The railroad constructed some buildings to house the section crew and their families. A school opened in 1910, and Edna Ross was the first teacher. The school operated until the 1920s. The railroad phased out the Pilot station in the 1920s. It moved the section crew elsewhere and dismantled the buildings for use at other stations.

From 1934 to 1938 a number of small mines, including the Badger, the Crazy Dutchman, the American Flag, and the Western Star remained active. Total production resulting from this bout of limited activity only amounted to 636 ounces of silver, 22 ounces of gold, and 833 ounces of copper. There was no other mining activity in the area until the 1980s, when Continental Lime took over the Pilot Limestone Quarry, which the Utah Construction and Mining Company had discovered but had not mined in 1960. The company completed a 400-ton lime plant, which is still in operation, in 1989. In 1990 it produced 103,000 tons of quick lime.

At Pilot, on the railroad, only a couple of concrete foundations and scattered debris mark the site. Little besides small ore dumps and collapsed workings remain at the mines, which are up Miners Canyon. Wagon ruts and rust marks on the rocks are evidence that sections of the California Trail once existed here. The Oregon-California Trails Association is currently marking the trail.

Tecoma

DIRECTIONS: *From Montello, head northeast on Nevada 223 for 7 miles. Exit right and follow the road for 1 mile to Tecoma.*

Tecoma was born upon the arrival of the Central Pacific Railroad to the site in 1869. A town, along with a small Chinatown, began to form during the next few years. A post office opened on December 1, 1871, closed on June 7, 1872, and reopened on May 2, 1873. Tecoma soon had a hotel, a restaurant, a cafe, and two saloons. By 1880 its population stood at sixty. The J.C. Lee store and a school opened in the 1880s. Tecoma became the major railhead for the area's mining activity, particularly that in Buel. The Sparks-Harrell Cattle Company also made extensive use of the town's rail facilities. In April 1886 S. Ross Worthington, a cattleman, and John Compton, a sheep

Foundations of the Tecoma depot. (Photo by Shawn Hall)

Residents in front of the J.C. Lee store in Tecoma, 1908. (Elizabeth Pruitt collection, Northeastern Nevada Museum)

man, became involved in an argument over Compton's dog at Mundell's Saloon. Both fired their guns and then died from the resultant wounds.

In January 1898 Lee sold his store to Jones and Johnson of Toano but later bought the store back and ran it until the late teens. A fatal railroad accident took place in Tecoma in August 1898. Jack Rouse, the superintendent of shops in Terrace, Utah, and machinist Peter Meaden were killed when the tackle broke around a derailed engine they were trying to right. By

1900 Tecoma had a population of 124. A big boost came in 1907, when the Southern Pacific Railroad built a spur from Tecoma to Buel. In addition, the Jackson mine, in the northern part of the mining district, began production.

During the next five years, mines around Tecoma produced more than $40,000 worth of primarily copper, but the town started to fade quickly in the 1920s. Its primary business, the hotel, had burned in 1918, and the post office closed on August 31, 1921. Mining slowed down, and Montello started to become the main rail station in the area. The completion of the Oregon Shortline in 1925 dealt another crippling blow to Tecoma, because until that time the town had been the main freight shipping point for Contact. By 1930 most of the town's residents had left, and all its businesses had closed. The railroad spur, which had not been used in years, was officially abandoned in January 1940. Over the years, the remaining buildings were dismantled for materials or moved elsewhere. Numerous concrete foundations and scattered debris mark the site, and a small cemetery is located nearby.

Toano

(Toana) (Summit City)

DIRECTIONS: *Located 1 mile west of Cobre.*

Toano came into existence as a stop on the Central Pacific Railroad in late 1868. The name is Shoshone for black-topped or black-coated, which the nearby mountains appeared to be. The new camp quickly achieved prominence when Central Pacific Railroad president Leland Stanford put together a special train at Toano. Headed by the engine Jupiter, the train was

John Cazier's hotel in Toano. (Northeastern Nevada Museum)

Grave in the Toano cemetery. (Photo by Shawn Hall)

featured at the golden spike ceremony—commemorating the completion of the first transcontinental railroad—at Promontory, Utah, in May 1869. Toano became the western terminus of the Salt Lake Division of the Central Pacific, and the railroad built a roundhouse with six stalls. A town formed and later became the major freighting and staging center in Elko County, outdoing even Elko for a while.

Toano was extremely important to the railroad, because it was not only the western terminus for the division but was also the first engine terminal east of Carlin. Toano quickly became the main supply point for towns as far away as Pioche, as well as for closer spots such as Spruce Mountain, Cherry Creek, and Ward. It also provided service to Idaho via the newly completed Toano Road. A post office opened there on August 9, 1869, and J. W. Griswold was postmaster. By 1870 Toano had a population of 117. Its businesses included the International and Railroad Hotels, L. E. Eno's Idaho Saloon, William Bellinger's saloon, the Nolan and Wilcox Saloon, and the Toano Saloon. A blacksmith, Charles Lynde, also worked in town. The post office closed on January 11, 1870, and reopened on January 10, 1872. In November 1871 Ferdinand Marx, the local Wells Fargo agent, completed a twenty-five- by sixty-four-foot fireproof stone store.

In December 1871 W. J. Bentley and James Chamberlain engaged in a fatal fight in the Toano Saloon. As the argument escalated, Bentley attacked Chamberlain, who was cutting tobacco. Chamberlain cut the artery in Bentley's arm, and Bentley quickly bled to death, but Chamberlain was acquitted of the murder. During the early 1870s a number of new stage lines made Toano their base. In 1872 Billings and Ellis began operating a stage from Toano to Tecoma and Buel. In 1873 Moffitt and Gossett started a twice-a-week stage from Toano to Cherry Creek, and Woodruff and Ennor started a tri-weekly stage to Schell Creek. In 1874 the Toano and Idaho fast-freight line began running. Critics said the line could not run because there was always mud in Thousand Springs Valley, but the company had few problems. The tri-weekly stage went through Contact on its way to Idaho. A school opened at Toano in August 1874, and Cecelia Hunter was the first teacher. Fires plagued Toano during its early years. On February 8, 1873, a large one destroyed the roundhouse and three engines. Some Chinese residents using fireworks started another fire on June 17, 1874. It destroyed the Railroad Hotel, and only residents' valiant efforts prevented the huge freight depot from burning.

Another death occurred in Toano in June 1881 when A. R. Smith, the railroad ticket agent, shot teamster William Nelson through the heart. Smith had the bad habit of pointing empty guns at people as a joke, and, unfortunately for Nelson, someone had left a loaded gun in Smith's office that day. The court decided that the shooting was accidental and released Smith.

While in 1880 Toano had a population of 123, the bottom was already beginning to drop out of the stage and freight market. The completion of the Oregon Shortline in the Snake River area in 1884 eliminated all the stage traffic heading to Idaho, and the town had to rely on the dwindling traffic heading south for survival. In 1887 John Cazier took over both Toano's store and its hotel. He ran them, as well as the post office, until 1898, when he sold his property and moved to Trout Creek in Starr Valley. In 1888 the Oasis Ranch Company began operations nearby. E. C. Hardy managed an operation that raised Norman-Percheron stallions, Polled Angus and Galloway cattle, and Spanish merino sheep.

The completion of the Lucin cutoff in 1904 was Toano's death sentence. The town was abandoned as a terminal point, and the repair depot closed. The Southern Pacific Railroad moved some of its buildings to the new division point at Bauvard, which later became known as Montello, and other buildings were torn down. Most of the town's businesses and residents moved. The few still there in 1906 relocated to the new town of Cobre, a mile to the east, on the newly completed Nevada Northern Railway. The post office closed on March 12 and reopened in Cobre. Cordelia Spencer was named the last postmaster when her predecessor, Leon Bednark, quit in August 1905 and left the mail lying in the street. The school also closed and moved to Cobre.

The town of Toano was dead by the summer of 1906. Many of its buildings were moved, along with its businesses, to Cobre. The Southern Pacific razed remaining structures when it rebuilt the tracks around Toano and added a second set of rails. Today, while no buildings are left at the site, there is plenty of evidence that there was once a town there. Huge stone foundations of the Marx store and the hotel are located in the center of many other foundations. Broken glass from the many saloons is everywhere. A cemetery containing more than thirty graves is located on the hill above town, though vandalism and occasional cloudbursts have left only a few legible markers, all on the graves of young children.

Wilkins
(Thousand Springs Valley)

DIRECTIONS: *Located 25½ miles north of Wells on U.S. 93.*

Long before Wilkins became a stop on the Oregon Shortline in 1925, the Thousand Springs Valley was an active ranching area. As early as November 1870 the railroad built a station in the valley to serve the Toano–Boise stage line. At times the valley became a quagmire, which made it extremely difficult for the heavy freight wagons to pass through. After the completion of the Idaho portion of the Oregon Shortline in 1884, there was no longer a need for the stage and freight line to Idaho, and it folded. A short-lived mining district, the Rough and Ready, formed in the valley in 1870, after some promising galena float came to light. The Mayflower, the Buttercup, the Good Friday, the Hillside, and the Elko mines also became active. Mines in the district, which was seven miles away from Wilkins, never produced anything, and the following year everyone in the district left. William Holbert relocated the mines in 1885, but once again they produced little, if anything. Around the same time John Bourne Jr. and John Bourne Sr. worked the Jane Day, the Mountain View, and the Hillside mines.

In 1886 John Sparks and John Tinnan consolidated many of the small ranches in the valley to create the base of their large ranching operation. Jasper Harrell had started the Winecup and the HD Ranches a few years earlier, but he sold his holdings to Sparks and Tinnan for $950,000 in 1886. Harrell used his profits to open a number of mines on Spruce Mountain. As

*Part of the Wilkins
Trading Post complex.
(Photo by Shawn Hall)*

*Water station and depot
at Wilkins. (Photo by
Shawn Hall)*

a result of the disastrous winter of 1890–1891, Tinnan sold his portion of the
cattle company to Andrew Harrell, Jasper's son, in 1891. The Sparks-Harrell
Company now controlled virtually all the ranches in the valley. Sparks later
sold his share to Harrell in order to pursue a career in politics. Sparks was
elected Nevada governor in 1902 and served until his death in 1908. Harrell

died in 1907. Ownership of the cattle company passed to the Vineyard Land and Livestock Company in 1908.

During the same time period B. B. "Bush" Wilkins, a former Wells bartender, began a ranch in the Thousand Springs Valley. In February 1911 he was sick in bed, when a drunken Joe Ford broke into his house and confronted him. Wilkins shot and killed the intruder, and the death was ruled as self-defense. The Utah Construction Company took over the Vineyard Company in 1921, and in 1925 the Oregon Shortline completed a branch to Wells and built a small depot and water station near the Wilkins Ranch. The siding became an important cattle-shipping point for the valley. Schools at the HD and Winecup ranches operated periodically from the teens to the 1930s. After the Utah construction mill folded, Russell Wilkins and Martin Wunderlich bought the holdings.

In the 1940s a hotel, a store, and a service station belonging to the Winecup Ranch opened on U.S. 93 near the ranch. John Moschetti managed the complex. The gas station was the first Phillips 66 station in Nevada, and Wilkins developed into a full-service truck stop with twelve employees, and a mechanic on duty twenty-four hours a day. The Wilkins post office opened on July 1, 1948, with Moschetti as postmaster. In 1950 Tom Bowers took over and enlarged the Thousand Springs Garage. Russell Wilkins died suddenly in 1952 while on a train heading to Salt Lake City, and much of his property was divided. Actor Jimmy Stewart and his partners Kirk Johnson, Arn Ehrheat, R. E. Harding, and Sue Harding Knott bought the Winecup Ranch and the complex at Wilkins for $700,000.

The Wilkins post office closed on April 12, 1963, when Moschetti left the Thousand Springs Trading Post. Over the years, he had operated the Wilkins businesses under a percentage agreement with whoever owned the Winecup Ranch at a given time. But this arrangement suffered when Bill Addington bought the property in 1962. After Moschetti had gone, the successful operation quickly deteriorated. The motel burned a few years later, and a fire damaged some of the other buildings shortly afterwards. Only the heavily vandalized shells of the buildings remain. At the Wilkins siding, which was abandoned in 1978 when the railroad folded, the depot/water station still stands amid some old corrals. The siding is a couple of miles west of U.S. 93. The Winecup and other ranches in the area continue to thrive. A kiosk on the road to the Winecup Ranch commemorates the California Trail, which ran through the valley, and gives a short history of the portion of the trail located in the area.

Southwestern Elko County

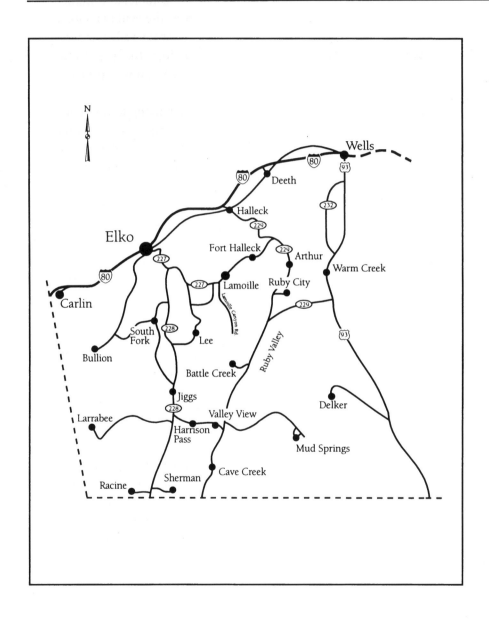

N

Wells

Deeth

Halleck

Elko

Fort Halleck

Arthur

Warm Creek

Carlin

Lamoille

Ruby City

South Fork

Lee

Bullion

Ruby Valley

Battle Creek

Delker

Jiggs

Larrabee

Valley View

Harrison Pass

Mud Springs

Racine

Sherman

Cave Creek

Lamoille Canyon Rd

Arthur

DIRECTIONS: *From Halleck, head south on Nevada 229 for 22 miles to Arthur.*

Isaac and Catherine Woolverton, along with their four children, were among the first settlers to arrive in the Arthur area. The Woolvertons and other families settled in Pole Canyon in 1868. By 1881 almost fifty people lived in the area, and nine students attended a crude one-room schoolhouse there. But the small town of Arthur did not come into existence until that name was given to its post office, which opened on April 21 with Arthur Gedney as postmaster.

There are two stories of how Arthur received its name. In one, the post office was named for Chester Arthur, who was U.S. president at the time, and in the other it was named for Gedney. The latter explanation seems more logical, since Gedney chose the name when he applied to serve as postmaster. Gedney originally homesteaded in the southern part of Ruby Valley. Besides his ranching interests, he ran a freight line from Wells to Eureka. He died on June 22, 1882, four days after a mare he had been riding kicked him in the head.

In 1884 a new school was built, but it soon proved inadequate. Len Wines took it upon himself to raise funds for a better one. The building was completed in early November 1886, and a celebration ball was held later in the month. Residents raised $31.50 to buy new school furniture and hired Emily Lemon as teacher.

For entertainment, balls and dances were held in Bachman Hall, located in Pole Canyon and presided over by Sam Bachman. During the 1880s Arthur was a stop on the Ruby Valley stage, run by Kelly and McCain. Nearby a Good Templars Lodge was organized in the early 1890s and was active for years.

In 1899 the Arthur school had five students. One, Oscar Griswold, later attended West Point, served with General "Black Jack" Pershing in World War I, and rose to the rank of three-star general in World War II. He served under Douglas MacArthur in the Philippines and was responsible for winning a number of decisive battles and liberating many concentration camps. Griswold, the oldest of ten children, was born near Arthur on October 22, 1886. His many responsibilities early on in life conditioned him for the rigors of the military. During Thanksgiving exercises at the Arthur school in 1899 the thirteen-year-old Griswold offered advice that he himself took to heart later during his military service. "Whatever we do, say or write," he proclaimed, "we should always BOIL IT DOWN!"[1] By the time he retired from the army in 1947 Griswold's awards for meritorious service included: the Army Distinguished Service Medal, the Navy Distinguished Service Medal, the Silver Star, the Bronze Star, the Air Medal, and the Legion of Merit. One of Elko County's "favorite sons," Griswold never forgot his roots and came back to Arthur and Ruby Valley whenever he could.

Ranching has always been the mainstay of North Ruby Valley. Children from the area's ranches kept the Arthur school running for years. Most ranches remained in the same family for many generations, and because the families were close to one another, many marriages occurred within the local population. In 1911 Margaret Griswold, sister of Oscar Griswold, married Jim Wright and moved back to the original Wright home at Arthur. They lived there until 1929, when they bought the Wes and Minnie Helth Ranch in Pole Canyon.

The Arthur post office continued to serve the ranching community until June 30, 1951. Many locals, including Jane Gedney, Jacob Wines, Andrew Hays, Leonard Covert, Erminie Woodhouse, Anna Woolverton, and Agnes Woolverton took their turns as postmaster. In July 1925 a fire destroyed the Isaac Woodhouse Ranch house. Postmaster Erminie Woodhouse and some brave neighbors saved all the mail and equipment before the fire burned the office.

A number of ranches continue to operate at and around Arthur. Many pre-turn-of-the-century buildings are still in use. Sam Bachman's hall has been replaced by Woodpecker Hall. Arthur residents continue to value the opportunity to get together, and they have preserved their strong sense of community and camaraderie.

*The Western Pacific
Railroad's Tobar depot
stood at Battle Creek
until burning in August
1995. (Photo by
Shawn Hall)*

Battle Creek

(Ruby Valley Mining District)

DIRECTIONS: *From Arthur, continue south for 24 miles. Exit right and
follow the road for 1 mile to Battle Creek.*

Battle Creek received its name in the 1870s, when Arthur Gedney
and William Myers, who shared irrigation water, disagreed over the amount
of water each should receive. The two stood toe to toe and battled it out with
shovels but later resolved their differences.

In 1903 John and A.M. Short discovered the Friday claim group, whose
ore contained silver, lead, and zinc. Soon afterwards, other prospectors began
to work the area. Witcher and Shaffer staked ten claims, and Andrew McDer-
mott also had claims. In 1906 the Friday Gold Mining Company bought the
Friday mine, and the Arthur Zinc Mining Company later took over the property. In July 1907 the Utah National Development Company began working
a group of claims, including the Friday mine, located near the head of Battle
Creek. D.C. Williams served as the mining superintendent. Residents constructed a few buildings, including a boardinghouse and a cafeteria, but the
company gave up in 1910 after producing about 100,000 pounds of lead and
400 ounces of silver. A copper deposit that produced over 4,000 pounds
during the next year before being mined out was discovered at Battle Creek.

In August 1917 Bachman, Blewett, and Woodward purchased the Michigan group, which consisted of six claims with values in lead and zinc, from
Frank Short. They constructed a fifty-ton coarse crushing mill, and a truck

hauled ore to Tobar at a cost of nine dollars per ton. A dozen men were employed in the operation, and during the next three years, 35,000 pounds of lead were produced.

From 1920 to 1942 only occasional leasing activity took place at Battle Creek. In 1943 Harvey Lambert of Ely and his associates discovered tungsten and scheelite in the Battle Creek mine. The mine was active until 1950, and it produced over three quarters of the district's overall output. From the time that Lambert et al. made their discoveries until the end of the decade, the mine produced 260,000 pounds of lead, 210,000 pounds of zinc, 3000 ounces of silver, 5000 pounds of copper, and 700 units of tungsten. In 1951 Jack White and his associates started the Noonday mine, located on the Friday claim group, on property purchased from Lambert, and Noonday Mines, Ltd. was organized. The mine was located at the mouth of Battle Creek Canyon. Buildings, including a boardinghouse at the Battle Creek mine one mile up the canyon, were moved to the new operation. In addition, the company purchased the old Western Pacific Railroad depot at Tobar. Leonard Hall moved it to Battle Creek, where he remodeled it as a boardinghouse and kitchen.

The owners spent the next year preparing the mine for production. A power plant supplied electricity to the mine, and plans were made to build a small mill. By 1953 the Noonday mine was producing and the mill had been completed, but the company enjoyed only three years of modest production. The mine produced more than 128,000 pounds of lead and 30,000 pounds of zinc, along with small amounts of copper, silver, tungsten, and gold. The company folded in 1956. Since then the area has seen only small amounts of activity. But in 1955 George Ogilvie and Lowell Thompson discovered a rich vein of tungsten in their mine. The vein yielded over 4,200 units of tungsten before it disappeared by the end of the year. That alone made the Battle Creek district the best tungsten producer in the county. There was a final flicker of interest in the area in the early 1980s when the Knight Roundy Mining Company developed the Battle Creek mine to produce tungsten, but the company left before any production occurred.

The Battle Creek site is now classified as a wilderness area, and the forest service is required, by mandate, to return the land to its original condition. In October 1993 the forest service confiscated the property, and the remaining buildings and mining equipment were slated to be auctioned or destroyed. One of the buildings left was the old Tobar depot. This author's plea to the Elko County commissioners prevented the depot's destruction, and efforts were underway to move the depot to Elko to become part of a railroad park and to serve the Elko Chamber of Commerce as an office. As one of the few remaining original Western Pacific passenger depots in Nevada, the structure was of extreme historical importance, but in August 1995 a large brush fire destroyed not only the station but all other buildings in the area.

Main street of Bullion, 1870s. (Mary Pisani collection, Northeastern Nevada Museum)

Bullion
(Bullion City) (Railroad City) (Highland) (Empire City) (Bunker Hill) (Railroad Mining District)

DIRECTIONS: *From Elko, follow Bullion Road for 4 miles. Head left and continue 20 miles to Bullion.*

Silver discovered in June 1869 on the east side of Bunker Hill created the first spark of interest in the Bullion District. Its proximity to the new town of Elko brought many prospectors to the district. In November of that year W. L. Grover discovered the Rainbow Ledge, and by the end of 1869 more than 200 claims had been filed. By the spring of 1870 the boom was in full swing, and two townsites developed. Highland was at the base of Bunker Hill, and Bullion City was platted about two miles down the canyon.

The biggest problem the Bullion District faced was the difficulty of smelting ore. In the spring of 1870 the Palisade Smelting and Mining Company brought in a furnace, but it proved a failure and shut down after three months. The company then built a new stone furnace, which also failed. By December both smelting works were under attachment for unpaid debts. On

June 22 the Rogers and Haskell furnace started treating Sweepstakes mine ore. It was the first smelter to have limited success. Most ore, however, still had to be shipped through Elko, and costs kept profits low. Nevertheless, the mines continued to produce. Active mines included the Bullion, the Ione, the Orphan Boy, the Brown, the Richmond, the California, the Mountain View, and the South Fork.

In the spring of 1870 George Shepherd completed the Bullion toll road, which originated from the Elko–White Pine Road at Wear's Station. J. F. Ray organized the Railroad Canyon toll road, J. F. Tasker opened the Railroad District toll road, and other stage lines began running to Carlin and Palisade. The 1870 census listed Bullion's population at 110, and by 1871 the area had attracted widespread interest. Highland contained a hotel, a store, a butcher shop, and an assay office, while Bullion City boasted a hotel, two saloons, one store, and twelve houses. Billings and Ellis began running a semiweekly stage to Elko via what is now known as Bullion Road. This lured business away from Shepherd's toll road. During the summer the Pluto Smelting Company installed a new furnace, which Adam Hay built, but like most smelters constructed in and near Bullion, it was a failure. A post office opened at Bullion City on August 8, 1871, with Joseph Phillips as postmaster.

More than fifty miners were now working the district, and mines were producing copper, silver, and lead. New mine discoveries included the Elko Tunnel, the Hussey Tunnel, the Our Kentucky, the Industry, the Miss Emma, the Wormer, the Last Chance, the Lee, the Lone, the True, the Red Jacket, the Tripoli, the Humboldt, the Webfoot, the Otto, the Republic, the Mayflower,

Overview of Bullion, 1870s. (Elko County Library collection, Northeastern Nevada Museum)

the Shoo-Fly, the Rhino, and the Bullion Extension. In August 1872 the Empire City Mining Company erected the first successful smelter in Bullion City. The smelter, which O. H. Hahn built, cost $30,000, had a capacity of fifteen tons per day, and treated ore from the Last Chance, the Elko Tunnel, and the Lone mines. A. J. Roulstone's eighteen-ton blast furnace, which cost $18,000 to build, started up on October 24. In 1872 production included 100,000 pounds of lead and $15,000 worth of silver, but it cost an average of $35 per ton to smelt the ore. New producing mines included the State of Maine, the Ella, the Wisconsin Chief, the Mountain Boy, and the Pine Creek. The town of Highland had evolved into a miners' camp, while Bullion City became the main town. Their combined population was about 150.

In February 1872 the Empire City Mining Company, which now controlled all smelting in the Bullion District, started a second smelter at Bullion City. The company also purchased most of the mines and claims in the district. Many of the small mines began to shut down as silver and copper veins pinched out, though two new copper mines, the Revere and the Bob Hunter, became active in 1873.

By 1874 Bullion had peaked. Businesses at Bullion City included the Bullion Hotel, which postmaster Joseph Phillips ran, the Billiard Saloon, and the Hoffman Boardinghouse. Only two main mining companies were active; they were the Empire City Mining Company and the Webfoot Mining Company. The Empire City Company worked the Lone, the Elko Tunnel, and the Empire City Tunnel mines. The Webfoot Company mines included the Webfoot, the Tripoli, and the Bullion Extension. In August a furnace, which J. B. Rand built, started up at the Tripoli mine. However, troubles appeared on the horizon. The Empire City Company folded in the fall, and the furnaces were dismantled, throwing about thirty men out of work. The Bullion District continued to lose residents, and by 1875 its population had dropped to less than fifty. Another severe blow came in 1876, when the Webfoot Mining Company suddenly folded and quickly left the district. The company owed miners over $10,000 in wages, and its debts to businesses in town amounted to even more. It already had a bad reputation for slow payment of debts. Despite much searching, company officials were never found.

Individual owners still worked a few of the mines, but the only mining company left at Bullion was the Lee Mining Company, which built a small smelter to treat its ore. In October the Mindeleff Smelting Company, financed by local mine owners, began to build a smelter that used a new process developed by a Russian scientist. The investors lost all their money and most of their property, when the project was exposed as scam. The company failed to pay its laborers before leaving the district.

After the economic disasters of 1874 and 1876, Bullion struggled to survive. Limited production still occurred, but the enthusiasm of the early 1870s had faded. By 1877 almost all mining in the area focused on copper. Mines still being worked included the Bullwhacker, the Grand Strike, the Blue Belle, the Ella, the Henry Clay, the California, and the Poor Man's Friend. In December a new furnace, which Edward Reilly built, started up. During the summer of 1878 Charles Brossman and John Norton reopened the Sweepstakes mine. In December George Bobier, who organized the Blue Bell Copper Mining Company, bought the Blue Belle mine and the Reilly furnace. In November 1879 the newly formed Elko Consolidated Mining and Smelting Company purchased the old Empire City Company holdings. By the summer of 1880 three mining companies controlled virtually all the mines. The new entrant was the Enterprise Mining Company, which owned the Webfoot. Long-time resident, A. J. Roulstone, was mine superintendent. Three furnaces were in operation at this time. Eliza Hoffman, who served as postmaster from 1878 to 1893, ran the hotel—one of Bullion's remaining businesses—which housed the post office. The school located across the street from the hotel opened in the 1870s, and fourteen students attended. Despite the many setbacks, according to the 1880 census, 150 resided at Bullion.

In January 1881 three new mines, the Bald Mountain Chief, the Tuolomne, and the Mono became active. In June Alfred Wild bought the Blue Belle mine and furnace and the Bald Mountain Chief mine, wich were producing eight tons of copper bullion a day. In July the Blue Belle Consolidated Mining Company was incorporated and paid Wild $30,000 for his holdings. The company worked the Blue Belle, the Bob Hunter, the Emma, and the Bald Mountain Chief mines and employed forty men. The company expanded in September, when it purchased the Tuolomne mine.

Bullion experienced another upswing in 1883. By summer 200 miners were living around the town, and two new saloons and a hotel were built. Board rates were ten dollars a week. The Bullion Milling and Mining Company began working the Lady of the Lake and the Silver Sides mines and built a five-stamp mill. George Pilz, formerly of Cornucopia, was superintendent, and he soon had the mill producing 300 pounds of bullion a week. Construction on a ten-stamp mill began in July and was completed on September 20. A twenty-ton-a-day furnace was also built. But the boom declined by 1884. The mills closed in 1885, and the furnace followed suit in late 1886. With the furnace's closure, the Bullion Company curtailed all operations. The value of its production amounted to $80,000. The Elko Consolidated Mines Company also folded in 1886, with the value of its production amounting to $120,000. Charlie White, a resident of Bullion for more than eleven years, sold his mine and moved to San Francisco in July.

The only producing mine in 1886 and 1887 was the Standing Elk, which had a total production value of more than $110,000. In 1887 H. K. Thurber of New York bought the mine from John Henderson, but not much happened after the sale, and by fall the mine was idle. In Bullion, population shrank to less than thirty and businesses closed. However, a sufficient number of children remained that the school stayed open. For the next thirteen years, the only activities to take place in the area were exploration and assessment work. Although reports exist of some small shipments of high grade ore, no production was recorded.

All was not quiet at Bullion, however. In November 1887 George Lewis, foreman of the Tripoli mine, murdered miner George Piccolli. There was bad blood between the two men, and Piccolli quit his job after they had an argument. Lewis would not let the disagreement die. He had a few drinks at the saloon, went home for his gun, and shot Piccolli three times at Picolli's cabin. Lewis spent the rest of his life in the Carson City state prison.

While there was plenty of ore available, the low price of copper and high cost of transportation made mining it unprofitable. Most of the work done during the 1890s involved stockpiling ore in anticipation of a price rise. In 1896 Fred Franks and John Mayhugh relocated the Ella mine and renamed it the Sylvania. By 1897 only twenty residents were left at Bullion. Franks, who had come to Bullion in 1873, was named justice of the peace. At this time, the Henderson Banking Company by default controlled most of the mines, including the Standing Elk and the Tripoli.

Activity began to pick up in the 1900, when five men started working the Brossemer mine, and Bullion's population rose to thirty-four. In December 1901 the newly discovered Raines Copper mine was bonded to Senator Clark of Montana. However, actual production did not occur until 1904, when 7,200 pounds of copper, 3,900 ounces of gold (more than the amount produced from 1869 to 1970), 15,000 pounds of lead, and 3,600 ounces of silver were mined. In June 1905 Charley Montgomery discovered a gold vein, and the property was bonded to the Amalgamated Mining Company of Butte, Montana, for $150,000. In August 1907 a copper strike took place in the Sweepstakes mine, which A. W. Hesson, Thomas Hunter, and J. J. Hylton owned. By 1908 a number of mines were active and the total value of production stood at $18,000. Three mining companies controlled the mines. They were the Delmas Copper Mining Company, the Nevada Bunker Hill Mining Company—of which J. A. McBride was president—and the Trimetal Mining Company—which owned the Copper Belle, the Copper King, and the Philippine mines. Without any mills or furnaces at Bullion, the ore had to be hauled by wagon to the railroad at Elko.

The Nevada Bunker Hill Company proved the most ambitious, spending the next eight years drilling a 7,000-foot tunnel in the Tripoli mine. By 1913 the Bullion District's revival was in full swing. The Nevada Bunker Hill Company contracted with Moe and Dougherty to build an aerial tramway from the Tripoli mine, located high on Bunker Hill, to the base of the mountain. The company improved the road to the Raines siding on the Eureka–Nevada Railroad, the preferred shipping route for most of the mines, because it was half the distance to Elko's railroad and much cheaper than hauling the ore the extra distance. In 1914 the railroad announced plans to build a fourteen-mile spur to Bullion. Bed was graded, but before construction began, trucks came into vogue, which negated the need for the spur. The value of district production for 1914 amounted to almost $100,000. A new company, which employed ten men, produced over half that amount. The Dome Mining and Reduction Company sent 1,700 tons of lead-silver-copper ore, which returned $57,000, to Salt Lake via Raines. However, 1916 proved to be the district's biggest year when it produced a record million pounds of copper. In addition, 700,000 pounds of lead and 65,000 ounces of silver were mined, for a total value of $343,000. Moe and Associates built a 150-ton copper smelting plant, which had a visible effect on the high copper production in 1916 and 1917. Producing mines were the Bunker Hill, the Lee, the Ennor, the Sweepstakes, the Sylvania, the Copper Belle, the Kentucky, the Hale and Peterson, the Grey Eagle, and the Bald Mountain Chief.

In 1917 production amounted to another 600,000 pounds of copper worth $200,000. While mines in the district produced every year until 1956, they yielded nothing close to 100,000 pounds of copper per year after 1917. The mill closed in 1918, and production of copper dropped to 74,000 pounds. Only in the years 1925 and 1950 was more than 70,000 pounds of copper produced again. In March 1918 William Patterson and J.D. McFarlane built a new smelter, hoping to benefit from the absence of treatment facilities, but the smelter was quite unsuccessful before it closed. This signaled the end of Bullion's rebirth, and by the end of the year the town's population had shrunk back to less than fifty.

Tragedy struck the Bett family, long-time residents of Bullion, when John Bett became the first Elko County casualty of World War I in March 1918. The Betts had been running a once-a-week stage to Elko and were involved in local mining. John's brother Jim took over the stage.

By 1920 the value of the district's production was just $3,500, and only the Bunker Hill and Sweepstakes mines produced. In 1920 the dormant Bullion school reopened for the eight children in town. Flo Reed, popular and well-traveled Elko County teacher, took over the school duties. She lived in the back of the Last Chance Saloon, which had an ornate back bar. It was the only business left in Bullion and an odd place for a teacher to live. Reed earned $125 per three-month semester.

The Frank family used the old Bullion store as their home, with the post office operating out of the living room. Mining continued to fade during the 1920s, and in 1925 the Nevada Bunker Hill Mining Company sold the slag heap at its old smelter and then folded in 1926. The reprocessing of the slag led to one last big production year, even though very little mining actually took place. More than 1,000,000 pounds of lead and 70,000 pounds of copper were recovered from the slag. From 1926 to 1949 mines yielded only small amounts of ore, causing Bullion to basically become a ghost town. The last vestiges of the town vanished in 1934 when on January 31 Fred Davis, who had been Bullion's postmaster since 1917, resigned. The office never re-opened. With only three students, the school also closed for good, and Jim Bett stopped running his auto stage.

Mining resumed from 1950 to 1956, but it did not revive Bullion. The Lead and Copper Mines worked the Aladdin mine (formerly the Standing Elk), and the Gregory brothers ran the Sweepstakes. During these six years, the mines produced 220,000 pounds of copper, 1,000,000 pounds of lead, 90,000 ounces of silver, and 350,000 pounds of zinc. An ambitious project, which involved digging a 6,000-foot tunnel to a target area, began in the Aladdin mine. But when the target was reached in 1959 there was no valuable ore, and the effort bankrupted the company.

During the next thirty years, much exploration but little production took place in the Bullion District. The discovery of Newmont's Rain deposit, just north of Bullion, has led to extensive exploration for disseminated gold. Newmont Gold, Westmont Mining, and Amax Exploration have all outlined deposits, and it seems only a question of time before active production begins. Leasing activity is still taking place on some of the old Bullion mines, but very little production has resulted. The total value of ore mined from the Bullion District is close to $5 million. Considering the relatively low-valued minerals produced, this is an incredible record. More than twenty-five million pounds of lead, seven million pounds of copper, one million ounces of silver, 440,000 pounds of zinc, and 35,000 ounces of gold have been produced.

Not much is left in Bullion today. Bullion City is marked by large slag heaps, smelter foundations, stone ruins, and a small cemetery that overlooks the town. At Bullion, which formed during the revival after the turn of the century, collapsed wooden buildings remain. A couple of buildings are standing, but they go back to more recent leasing operations. Highland, located near Bunker Hill about two miles from Bullion City, is marked by extensive mine dumps, stone ruins, and mine tunnels.

Cave Creek
(Cave City) (Shantytown)

DIRECTIONS: *From Arthur, continue south for 52 miles to Cave Creek.*

A group of soldiers from Fort Ruby named Cave Creek, when they discovered a cave entrance at the creek head. After they made their discovery, the soldiers rowed into the cave to explore, but the effort turned disastrous when the boat capsized, and all but one of them drowned. In 1963 Mildred Breedlove, Nevada's poet laureate, wrote a poem memorializing the incident:

Deep in the Rubies, and almost unknown,
Where the country is rugged and wild,
Is a stream flowing out of a cave.
Low on the mountain, where boulders are piled,
Flowing water is polishing stone
At the mouth of a watery grave.
Fearless explorers went in with their boats,
And the underground lake that they found,
Was as deep and as dark as the night.
Walls of the cavern withdrew, while the sound
Of imprisoned, unmusical notes
Was a devil that laughed at their plight.
One member returned, with an angel as Guide,
But the horrors had taken their toll,
And the cavern refused to be crossed.
Stricken and gray, he had learned, in the hole,
Many secrets he dared not confide,
But the one without guidance was lost.

Further exploration took place at Cave Creek a few years later, and the "Great Organ" formation was discovered. There are differing accounts of who the discoverers were. One report gives credit to A. G. Dawley and Tom Short, and another cites T. S. Dawley and Tim Haley. When Short explored the cave, he found the three soldiers' bodies.

Because of its proximity to Fort Ruby, a small camp had formed at Cave Creek by 1867. Samuel Woodward and Michael Flynn operated a distillery that produced whiskey called Old Commissary. This made Cave Creek popular with soldiers. Woodward and Chester Griswold built a sawmill and ran a restaurant and saloon, but these enterprises continued to exist only while Fort Ruby was active.

Patronizing Cave Creek's saloon was not the soldiers' only pastime. They combed the mountains above Cave Creek and discovered silver ore. The Cave Creek Mining District organized in May 1869, and active mines included the Amazon, the Mississippi, the Red Pine, the Murphy, the Dodd, the Enterprise, the Exchequer, and the Longmore. As a result of this activity Cave City formed, but the mines produced little before they were abandoned in the 1870s.

Once Fort Ruby closed, ranching was the only business venture being pursued at Cave Creek. Thomas Short purchased the Cave Creek Ranch from George Williamson in 1872. Short, who was born in Ireland and who married Mary Devine, was one of the first cattlemen in Nevada to register his brands. He served with Sherman during the Civil War and was one of the original locators of the Richmond mine in Eureka. In 1882 he sold the Cave Creek

Ranch to Jacob Bressman and bought the Thompson Ranch, located ten miles to the north. It became the Short home ranch. Short specialized in breeding purebred shorthorn cattle which won blue ribbons at the International State Show of 1910 in Chicago. His wife died in 1913, and he never recovered from the loss. He died in August 1918 and left seven children.

The Cave Creek Ranch remained the property of Jacob Bressman until his death in 1896. While he owned the ranch, a post office opened on November 5, 1887, with Catherine Flynn as postmaster. The office remained open until April 30, 1929. After Bressman died, the ranch passed to Lou Benson, who had been a very popular teamster on stage lines to Eureka and Hamilton. Benson ran the ranch until he died in July 1927, when Albert Hankins took over. He then sold the ranch and nearby marshlands to the government for the purpose of establishing the Ruby Lake Migratory Water Fowl and Game Refuge. In addition, forty acres were set aside for a fish hatchery to replace the closed facility on Trout Creek. Elko County ran the Ruby Valley Fish Hatchery until 1947, when it became a state and county entity, stocking fingerlings in creeks and lakes throughout Nevada. The state officially purchased the facility in 1963. On August 26, 1967, its name was changed to the Dr. Harry M. Gallagher Fish Hatchery. At its peak, the hatchery produced over 200,000 pounds of fingerlings a year. The facility continues to operate today and is one of the most productive hatcheries in the state.

Remaining buildings from the Cave Creek Ranch are now the base for the Ruby Lake National Wildlife Refuge, one of the most popular recreation sites in Elko County. The Cave Creek school, which served Southern Ruby Valley for years until the 1970s, still stands. A private cemetery at Cave Creek contains a number of graves, including Jacob Bressman's. A campground is located nearby, and limited supplies are available at Shantytown.

Deeth

DIRECTIONS: *From Wells, head west on I-80 for 18 miles. Exit south and continue 1 mile. Turn left for ½ mile to Deeth.*

Before whites settled at Deeth, the area was used as a seasonal Shoshone camp. Beginning in the 1840s emigrants began flowing past the future Deeth townsite on the California Trail. There are two stories about the origins of the name Deeth. In one, an enterprising man named Deeth opened a small store and trading post on the banks of the Humboldt River about two miles away from Deeth. In another, early emigrants called the site "Death," because travelers without enough water would meet their death in the desert. As the story has it, "Death" later became "Deeth." Early journals from the 1840s talk about a place called Death to the west of Humboldt Wells (now

Wells). Because the trail was so heavily traveled, no settlement developed, though thousands of emigrants passed through.

Deeth formed in 1869 when the Central Pacific Railroad was completed in the area. As construction continued eastward to Promontory, Utah, the railroad built a sidetrack station complete with a telegraph line. At first just a few people lived at Deeth, and then a town slowly began to form. A post office, which was located in a boxcar that was also the railroad depot, opened on November 2, 1875, with John Donne as postmaster. That same year, Tom Atkinson opened a saloon, the first business in Deeth. In 1877 James Porter built a large two-story building that served as a store and hotel, and his brother, John, opened a livery business. Atkinson sold his saloon in 1879 to Ed Seitz, one of the first Pleasant Valley settlers. Seitz also served as an Elko County sheriff. He immediately built a hotel addition to his saloon.

Deeth's population rose to thirty-one by 1880, and the railroad built a warehouse and a water tank and added a section crew. Deeth became the supply center and shipping point for the Starr and Ruby valleys, and a stage line to Mardis (otherwise known as Charleston) ran three times a week. In

Southern Pacific depot at Deeth. (Northeastern Nevada Museum)

Southern Pacific depot at Deeth. (Northeastern Nevada Museum)

1887 C.C. Truett built another hotel in Deeth. William Mayer, who would eventually run the Mayer Hotel in Elko, later purchased Truett's hotel. The J.R. Bradley Cattle Company gave Deeth's growth a big boost, as did Bradley's presence itself. Bradley opened a store in Deeth and later another at Mardis. His company competed directly with Porter, and Bradley's newspaper ads listed Porter's prices and his own lower ones. The spirited competition had Deeth residents smiling because price wars forced prices as low as those in nearby Elko.

Literary enlightenment came to Deeth in 1896, when the Reverend Merchant Riddle organized the *Deeth Tidings*. Riddle was sole owner and used the paper to promote the Free Silver Party platform. The paper was an eight-page weekly with a subscription rate of two dollars per year. Members of the local Good Templars Lodge served as editors. The townspeople of Deeth relished their local paper, but the population was too small to make it viable. In September Riddle moved his newspaper plant to Elko and started the *Nevada Silver Tidings,* which ran until July 1899.

In 1899 Porter gave up his battle with Bradley and sold his store, eight-room hotel, home, corrals, and stables to A.C. Dorsey. A school was built in 1899, with Isora Stevens as the teacher; thirty students attended in the first year. By 1902 William Mayer and Oliver McCall were running the Dorsey store and hotel. In September 1903 a Chinese cook who was smoking opium in his room accidentally set Seitz's old hotel, then owned by Charles Standley, on fire. The hotel was a total loss.

By 1910 Deeth was approaching its peak. New strikes at Jarbidge sparked a great revival of the stage and freighting business. With some financial backing from the Southern Pacific Railroad, local businessmen and ranchers raised $10,000 to construct a stage road to Jarbidge in an attempt to lure business away from Jarbidge's only other source of supplies, Twin Falls, Idaho. The effort paid off, and except when snow forced closure of the road, Deeth enjoyed a lucrative trade. Fare by stage from Deeth to Jarbidge was ten dollars, and ten-horse teams left daily for Jarbidge. In 1910 the Western Pacific Railroad was completed through Deeth, and the railroad built a large depot. C. M. Haws surveyed the town in December 1910, and the township of Deeth was established. Deeth endured a major flood in 1910, when ice piled up against the Western Pacific bridge just east of town. The ice dam forced the Humboldt River to flow through the town, but luckily damage was slight.

New businesses continued to open, including the Deeth Mercantile Company, which was housed in a 4,000-square-foot building. A. B. Gray and E. C. Riddell built the Deeth Creamery, which opened in May 1912. It produced 200 pounds of butter per week but was not very successful and closed within a year. The facility later moved to Trout Creek at Welcome and was used as a fish hatchery until it was abandoned, when the Ruby Marsh hatchery opened. Gray brought other interests to Deeth, including his *Commonwealth* newspaper, which he moved from Carlin to Deeth in February 1912. The paper was a democratic weekly, and he published it until October 1914, when he moved and purchased the *Carson Weekly* in Carson City.

In 1912 Oliver McCall, part owner of a local store and hotel, won the mail contract to Jarbidge. He set up stations at Pole Canyon and Hank Creek. McCall built a strong reputation, keeping to his twice-a-week delivery schedule even during the tough winters. In the early teens, Deeth's population peaked at 250. There were two stores, two hotels, a dance hall, the Bradley Opera House, five saloons, a slaughterhouse, a red-light district, the Cannon Pool Hall, and twenty-five homes. In July 1913 the Mahoney Brothers saloon and the Nichols's saloon burned at a loss of $10,000. This began a trend of devastating fires at Deeth. The Southern Pacific and Western Pacific Railroads both had large depots in town, but the opera house was the town's pride. Many touring troupes performed, as did the Deeth Orchestra. While Elko preachers occasionally held religious services in Deeth, a church was never built.

On January 22, 1915, John Seabury started a new weekly newspaper, the *Deeth Divide*. However, reports make no further mention of the paper, and it apparently folded shortly after it began coming out. A huge fire, which started in the Mayer Hotel, struck Deeth in October 1915. Before it had burned out, two thirds of the business district was destroyed. Businesses that burned out included the Nicely Store, the Truett Saloon, the post office, and the rebuilt Mahoney Saloon. While the exact cause of the fire was not established, there were suspicions that the financially strapped hotel owners had set it to collect insurance money. The fire effectively ended Deeth's prosperous times for good.

The arrival of the automobile further decreased the need for Deeth's remaining businesses, and during the ensuing years many buildings and homes moved to Wells, Elko, and other locations. Another fire in the 1930s claimed the Bradley Opera House, which had been converted to a hotel, and the old Smiley store, home of the post office. The Deeth school closed in 1957 after it consolidated with the Wells School District, and in the 1950s the Western Pacific depot moved to Crested Acres, where it is still used as a restaurant and bar.

Deeth still has a population of about twenty, but only a handful of original buildings survive. The Deeth post office remains open and is housed in a cinder block building. Deeth was home to a number of prominent citizens over the years. Two governors, Bradley and Russell, came from Deeth. State legislators from Deeth have included James Riddell, Ebenezar Riddell, David Johnston, Joseph Johnston, Jim Russell, Edward Murphy, Morley Murphy, and Hugh McMullen. Stanford Weathers was a West Point graduate, and Arthur St. Clair was a Rhodes Scholar.

Delker

(Delcer) (Locust Spring)

DIRECTIONS: *From Currie, head north on U.S. 93 for 4½ miles. Exit left and follow the road for 14 miles. Turn left and follow that road for 2 miles to Delker.*

In 1890 a man named Delcer discovered a gold prospect at the Franklin Buttes, now called the Delcer Buttes. A mining district was organized, and because of a filing error the district was called Delker. Only two mines, the Delker (also known as the Delcer or the Copper Belt) and the Emma, were ever worked, and they produced little.

Discovery of the Delker in 1894 was followed by the Emma's discovery in March 1899. In August 1907 Ella Rawls Reader, one of the backers of the proposed Wells-Bald Mountain Railroad, paid $10,000 to purchase the claims of J. H. Stratton and Charles Marley, which covered 320 acres. Despite Reader's efforts, the ore was not valuable enough to ship. The district's only recorded production occurred in 1916–1917 when 98,000 pounds of copper, shipped to Utah via Currie, came from the Copper Belt mine. J. H. Stratton of Cherry Creek and William Griswold of Ruby Valley owned the mine. Unfortunately, the ore was located on a badly faulted vein, which disappeared in 1917. Since 1917 no production has occurred at Delker, though the Gold

Creek Company of Ely, the Exxon Minerals Company, and the Pegasus Gold Company did exploratory drilling in the 1980s. Only a shallow shaft and ore dumps mark the Delker site.

Fort Halleck
(Camp Halleck)

DIRECTIONS: From Halleck, head south on Nevada 229 for 11 miles. Exit right and follow the road for 6 miles to Fort Halleck.

Fort Halleck came into being on July 26, 1867, when Captain Samuel P. Smith, Lieutenant Augustus Starr (Starr Valley's namesake), and seventy men arrived from the Winnemucca area and established an army camp there, as they had been ordered to do on June 30. The troops had to live in dugouts until a proper barracks was built in the fall, and then in tents until their quarters were completed in February.

Smith had been in the Elko County area before, when he served at Fort Ruby, and after he left the military he ran a business in Elko for awhile. He was reputed to be a ruthless man. Fifteen of the seventy men who had originally come with him to Fort Halleck deserted rather than face his severe brand of discipline.

Camp Halleck was named for Major General Henry Halleck, commander of the U.S. Army. In May 1868 Fort Churchill closed, and all military personnel were transferred to Camp Halleck. A post office, named Camp Halleck, then located in Lander County, opened on October 21, 1868, with J. P. Olds as postmaster. Despite this activity, the quality of life at Camp Halleck remained poor until the Central Pacific Railroad was completed through Halleck, twelve miles to the north, in 1869. With the railhead so near, lumber, brick, and supplies began to flow into the camp. New buildings were erected, and the camp began to look more like a fort. A double murder occurred near the military post on February 29, 1869, when an undiscovered culprit killed two Jewish peddlers with an ax and then stole their goods. The case was never solved. These were the first murders in what was to become Elko County in March.

The 1870 census put Camp Halleck's population at 145. By then, the camp had become the social center for many of the nearby valleys. It hosted military balls, and the railroad ran special trains from Elko, complete with sleeping cars. The post's twenty-four-piece military band provided the music for the balls held at the camp's hall. John Day offered additional entertainment at his nearby ranch. The camp's soldiers were frequent patrons of his popular saloon and dance hall. In October 1870 Edward Carr, a soldier with Company One of the Third Cavalry, imbibed a little too much at Day's dance

Soldiers drilling at Fort Halleck, 1870s. (National Archives collection, Northeastern Nevada Museum)

hall, fired a carbine at a sergeant, and hit a dance-hall girl, fatally wounding her. Because the incident happened off Camp Halleck's grounds, Carr was arrested and sent to state prison.

To escape the tedium of camp life, a group of soldiers formed the Camp Halleck Variety Troupe in 1871. J. M. Wood was the manager, E. Baker the stage manager, J. D. McClellan the secretary, and Frank Stanton the head of the orchestra. The troupe entertained not only at the camp but elsewhere in Elko County. It received a black mark when in August one of its members, Thomas Carter, attempted to rape Mary Brown, while the troupe was visiting Elko. Mary's husband rescued her after he heard her cries for help. Carter was arrested and sentenced to ten years in prison.

Because of the camp's size, the demand for goods, most of which local ranchers provided, was high. One of the most prominent ranchers in the area was John Dorsey, who had settled near Camp Halleck in 1870. Previously, he had run a store, a Wells Fargo station, and a wood-choppers camp, all at Cedar Pass on the Central Pacific Railroad. After moving to Camp Halleck, he raised some of the best purebred shorthorn cattle in Nevada. They commanded prices as high as $20,000 each. His wife, Eliza Lyon, had been a member of the Wilmington Overland wagon train in 1850. The couple married at Diamond Springs, California, in 1854. Eliza was one of the first women in Elko county with a brand registered in her name. John was a member of the first Nevada State Cattle Association in 1884. Eliza died in January 1893, and John followed in April 1897. Charles Mayer, who later built the Mayer Hotel in Elko, also got his start at Camp Halleck, where he became post trader in 1872. Jack Abel, owner of another nearby ranch, ran a blacksmith shop and, more importantly for the soldiers, a saloon that provided live entertainment—including comedy troupes and music.

All these diversions cost money. Military personnel were paid in government greenbacks, but to buy goods or a drink required gold and silver, and there was a standard 50 percent premium for converting the greenbacks. Prices for everyday supplies were quite high. A dozen eggs cost $2 in gold, and butter went for $2.50 a pound.

Beginning soon after the completion of the Central Pacific at Halleck and continuing until soldiers abandoned the camp in 1886, there was constant pressure in Washington, D.C. on the government to move the fort to the railroad. However, the fort had cost so much to construct that the government never made the move. In 1873 fire struck the camp and destroyed some buildings, including one that housed the mining locations book for the Halleck Mining District. This caused great confusion, because all the papers had to be refiled. By the mid-1870s Camp Halleck had grown to over twenty buildings, including a hospital, officers' quarters, enlisted men's barracks, a commissary, a magazine, a granary, stables, a guardhouse, a mess hall, a bakery, and a blacksmith shop. A school opened in 1875, with the post chaplain serving as teacher of both military personnel's and civilians' children.

In April 1879 Camp Halleck was officially renamed Fort Halleck. As a result, the post office changed its name on May 17, 1880. During the early 1880s pressure mounted to close the fort, because it was no longer a sensible venture. One of the two companies at the fort, the First Cavalry, was transferred to the Dakotas in 1884. The inevitable closure was approaching. In February 1885 Mrs. G. R. Kemp, a soldier's wife, started a monthly paper, the *Halleck Gossip,* designed to lift the spirits of those remaining at the doomed fort, but it apparently only lasted a short time. In 1886 Major General O.O. Howard reported that, "there seems to be no good reason why Fort Halleck, Nevada, should not be abandoned. It is 12 miles from the railroad and possesses no paramount importance as a strategic point. The settlers are interested in some degree in keeping up the Post, in order to have a market near at hand for grain and other supplies that they can raise. It is, considering its size, the most expensive Post in the Department." [2]

The post's abandonment became official on October 11, and the First Infantry was transferred to Fort McDermitt. After the official closure on December 1, the fort was empty except for caretakers. Residents of the surrounding area pilfered much of the fort, including a wooden fence built to keep out looters. One such looter apparently accidentally burned down the hospital in July 1890. An investigation to arrest those responsible for looting the fort began but ceased abruptly when everything at Fort Halleck was sold at auction on February 1, 1898. Isaac Woolverton bought the commander's large and beautiful home for $250, and the schoolhouse was sold for $55.

Ed Helth bought the storehouses for $6, Chris Lund paid $10 for the power-house, and Edward Kelley purchased the liberty pole for $8. Soon afterwards, all buildings except the schoolhouse had been removed. The school continued to operate for a number of years amid the ruins of Fort Halleck. The post office remained open until May 15, 1907, serving mainly local ranches.

There were quite a few graves in Fort Halleck's cemetery, where both military personnel and civilians were buried. Most date back to the town's early years, when conditions were very poor. The soldiers' remains were moved to the National Cemetery at the Presidio in San Francisco in 1887, and only a few marked graves remain.

The Fort Halleck site is marked by a grove of huge trees planted in the 1870s. Foundations suggest the general layout of the fort, but not much else remains. The Daughters of the Utah Pioneers placed a monument along the road in 1939. The only buildings left from the fort are located on the grounds of some nearby ranches.

Halleck

DIRECTIONS: *Located 19½ miles east of Elko, 1 mile south of I-80.*

Halleck came into existence in 1869 when the Central Pacific Railroad was completed through the future townsite, which immediately became the shipping point for supplies heading to Fort Halleck. In 1869 two hotels, the Bell and the Griffin, opened, and Frank Hughes constructed the town's first home, which was made of adobe. By 1870 Halleck had a population of thirty-five. In the early 1870s the town experienced a building boom of sorts when F.T. Greenberg opened a brewery and Hamilton McCain built a saloon catering to the soldiers and officers of Fort Halleck. In 1873 Bell, the Halleck depot operator, expanded his Halleck Hotel to make it two stories, including a second-story porch, but horses tied to the support columns tugged and loosened the supports so much that the porch eventually collapsed. Sam Moser opened a store, and a school was built in 1874. The Halleck School District, sometimes referred to as the Peko School District (a school by that name also existed in Elburz), remained open until the 1950s. A post office opened on April 24, 1873, with Henry Hoganson as postmaster, and Hamilton McCain built a two-story hotel of his own adjoining his popular saloon.

Halleck developed into the heaviest livestock shipping point on the Central Pacific. In the mid-1870s Bell sold the Halleck Hotel to John Deering, but Bell stayed in Halleck, where he served as postmaster and constable. By 1875 Halleck's permanent population had grown to fifty, though there were always more people than that in town. A tragic event occurred in Halleck on

Western Pacific depot at Halleck. (Lois MacKay collection, Northeastern Nevada Museum)

Halleck Hotel, 1907. (Northeastern Nevada Museum)

April 9, 1877, when Sam Mills, whom Mrs. Deering had fired for poor work at the Halleck Hotel, reacted angrily. When another man knocked him down to protect Mrs. Deering, Mills pulled a knife and then went to the Griffin Saloon where he got a gun. Mills's friend, James Finnerty, wanted to talk to him, but as soon as Finnerty opened the door, Mills shot him dead. Mills fled and was later captured in a haystack near Lamoille. He was found guilty of first-degree murder. Mills vigorously objected to the verdict, condemning the jury for its decision, because he had never intended to kill his best friend.

On December 22 Mills hanged in Elko. His was the first legal execution in Elko County.

In the ensuing years Halleck's population gradually grew. By 1880 it stood at 97 and by 1900 had risen to 126. At the town's peak, twenty-six buildings stood, and numerous businesses flourished. But Fort Halleck's abandonment in 1886 ended Halleck's glory years, and the town had to depend on local ranchers for survival. Ranching has been and still is an important element of the Halleck area. One of the most famous ranchers of the time was Dan Murphy. Murphy and his father, four brothers, and four sisters were part of the Stevens-Murphy-Townsend Party, the first emigrant wagon train to come to California in 1842. He was also the first white man to see Lake Tahoe. While the bulk of Murphy's cattle empire in Elko County was in the North Fork area, his base of operations was in the Griffin Hotel, and in 1877 he built a home of his own in Halleck. When Murphy died in 1882, he owned four million acres of land, making him one of the landowners with the largest landholdings in the world.

Samuel McIntyre was another prominent local cattleman. Phillip Witcher had established a ranch in 1880, and McIntyre purchased it in 1888 and ran 15,000 head of Black Galloway cattle imported from Scotland. Because his cattle were of a unique type, he did not have to brand them. McIntyre was also interested in mining and discovered the Mammoth mine in Utah. In 1898 he added the Brennen and Dorsey ranches to his holdings. Surprisingly, many of his holdings, which Frank and Phyliss Hooper now own, are still intact. The Hoopers use the original McIntyre Ranch as their headquarters.

The 71 Ranch, which Joe Scott established in 1877, is one of the largest ranches in the area. Caleb Hank joined with Scott in a partnership in 1879, and in 1889 the two formed the Halleck Cattle Company. The ranch prospered until William and Grace Large bought the property at public auction in 1911 for $170,000. The Larges formed the 71 Ranch Corporation in 1913. The Union Land and Cattle Company took over the corporation in 1917 but folded in 1921 when cattle prices plummeted. John Marble ended up buying the ranch, and three generations later the 71 Ranch was still in the Marble family. The Ellison Ranching Corporation bought it in 1995.

In the 1890s the Halleck baseball team entertained residents. But the town was slowly dying, and many of the buildings were moved or torn down. By the turn of the century, the Halleck Hotel was the only substantial building left in Halleck. Stanley Wines ran the hotel and store and also had the mail contract between Halleck and Ruby Valley. He sold the hotel in 1913 to Samuel McIntyre, but by the time the Halleck Hotel burned in 1915 there were only a couple of buildings left in town. McIntyre vowed to rebuild the hotel, but later decided it was not worth the investment.

The Halleck post office continues to operate from one of the two remaining buildings in town, and it serves the many local ranches. For many years, the post office had a mascot named Little Dog. The dog had lived in the back room of the post office when Earl Conrad was postmaster. Conrad died in 1966, but the dog stayed, and Nelda Glaser and post office patrons took care of him. In 1972 a pack of dogs attacked Little Dog, and the small dog sustained severe injuries. Dr. Cuthbertson of Elko operated on Little Dog and noted that the three-hour surgery was the longest he had ever performed. While Little Dog recovered, post office patrons brought him extra food and scraps. Many people, including a former resident of Halleck living in Nampa, Idaho, who had heard about the attack, contributed towards the dog's vet bill. Little Dog remained a fixture at the Halleck post office until he died a few years later.

Harrison Pass

DIRECTIONS: *From Cave Creek, head north for 12 miles. Exit left and follow the road for 1½ miles to Harrison Pass.*

Thomas Harrison, a native of England, first settled Harrison Pass in 1865, when he established a 1,500-acre ranch there. Initial ore discoveries in Harrison Pass were made in January 1897, when a four-foot ledge bearing a high percentage of copper, two and a half ounces of silver, and gold worth ten dollars per ton began producing. However, little more development occurred before the claim was abandoned. It was not until 1916 when Russell Campbell discovered a scheelite deposit—the Harrison Pass

Camp and mill of the Star Tungsten Mine. (Edna Patterson collection, Northeastern Nevada Museum)

A number of buildings remain at the mining camp in Harrison Pass. (Photo by Shawn Hall)

mine—that interest in mining revived. While the deposit developed to a limited degree, no production was ever recorded. Campbell, who continued prospecting the Harrison Pass area for many years, discovered a tungsten deposit—the Campbell mine—in October 1929. This discovery led to a period of exploration and production. By 1940 the Star Tungsten (located at the site of the old Harrison Pass mine), the Lakeview, the Slipper, the Climax, and the Campbell mines were being developed. A small mill started operations at the Star Tungsten mine, which George Ogilvie, Ed Lane, and Andy Francis owned, but a fire destroyed it in February 1943. The owners sustained a $25,000 loss, but they, nevertheless, rebuilt the mill in 1944. Soon after its completion, however, operations ceased. The mine closed in July, and the machinery was removed from the newly completed mill.

Since then the district has seen only minor levels of production in the mid-1950s and late 1970s. In total, it has produced 15,000 units of tungsten. In the 1940s a small camp that developed near the mill supported a school. Today, the mill foundations are still visible, and the mine dumps of the Star Tungsten and the Climax mines are nearby. A number of occupied cabins exist at the camp.

Jiggs

(Cottonwood) (Dry Creek) (Mound Valley)
(Skelton) (Hylton)

DIRECTIONS: *From Elko, head east on Nevada 227 for 3.7 miles. Take Nevada 228 south for 26.8 miles to Jiggs.*

Although Jiggs was never large, its history has many unique facets. During its lifetime, six different post offices have served the town, and it achieved national prominence when Volkswagen put all its residents in one of its vans for a nationwide advertising promotion.

W. M. Kennedy, the first settler in Mound Valley, arrived in 1866, though Myron Angel, in *History of Nevada,* erroneously identifies the year as 1861. Kennedy named the valley for a large mound located near his property on Smith Creek. By the late 1860s there were quite a few homesteads in the valley. The boom at Hamilton (in White Pine County) led to the formation of a number of toll roads, including the Hill Beachey, the Gilson Turnpike, the Woodruff and Ennor, the White Pine toll road, and the Eureka–Hamilton road, which all came through Mound Valley. The main stop in the area was Hooten's Station, located east of the present Jiggs townsite.

The first post office in the Jiggs area was named Cottonwood. It opened at the Porter Ranch on December 14, 1869, with Mason Dexter as postmaster, and closed on July 12, 1870. Cottonwood was home to a strange mixture of hard-working, law abiding residents and wild outlaws who called the camp at Hooten's home. The latter group's reputation inspired popular western author Zane Grey to make Cottonwood home base for his fictional outlaw King Fisher.

But as the 1870s progressed, the area became a respectable ranching community. A new post office, Dry Creek, opened on February 24, 1874, at the Porch Ranch, with Augustus Miller as postmaster. Henry Porch took over the job in 1876 and served until the office closed on March 26, 1879. Porch, who settled on Dry Creek in 1872, ran a way station on the stage line and built a log cabin to house the post office. In 1877 his brother, Thomas, bought the land and moved his wife, Martha Ricketts, and their six children to Dry Creek. He and his wife then ran the station, providing overnight lodging and meals to travelers heading to Hamilton. Six more babies were born at the ranch, and Thomas taught the twelve children since there was no local school. After he died in 1893 Martha continued to run the station and dairy business until she sold out in 1910.

The day the Dry Creek post office closed, the Mound Valley office, with David Hooten as postmaster, opened at Hooten's Station. While the Mound Valley office closed on March 17, 1881, Valley Paddock, a former telegraph

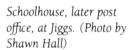

Schoolhouse, later post office, at Jiggs. (Photo by Shawn Hall)

operator for the Central Pacific Railroad, petitioned to reopen it in 1884 under the name Skelton—his mother's maiden name. One historical source says Skelton was named for a local evangelist, but the claim is unsupported. The Skelton post office was open from November 21, 1884, until May 1, 1911, when the name was changed to Hylton—in honor of prominent local businessman John Jesse Hylton. The new office closed on November 30, 1913. The present Jiggs office opened on December 18, 1918.

While the town of Jiggs has existed for 125 years, the real impetus of its development was the arrival of John Jesse Hylton in 1874. After working on Mound Valley ranches for years, he bought the Cedar Hill Ranch, near Harrison Pass, in 1880. This was the first step towards the construction of a huge ranching empire. In 1889 he married Lena Garrecht, who mother ran the Elko Hot Springs Hotel. After the devastating winter of 1889–1890 wiped out most of Hylton's sheep herd, he decided to try the cattle business. He purchased the Miller Ranch on Huntington Creek from Ed Carville. It became known as the Hylton Home Ranch. Soon afterwards, Hylton also bought the Hale and 25 ranches, and in 1900 opened a general store with prices lower than those in Elko stores. The location was excellent because it was much more convenient for large sheep operators to shop there than to go to Elko for supplies. It did not hurt that Hylton was a sheep man, either. At the same time, Hylton brought phone service to Jiggs, forming the Elko Southern Telephone Company. The switchboard was located at the Circle Bar, run by Colon Hylton, John's sister-in-law. Because of demands on his time, John Hylton hired his brother-in-law, George Hanna, to run the store, which became known as Hylton and Hanna General Merchandise. However, the arrival of the automobile decreased the demand for a local store, and Hylton sold out to Albert Hankins in 1916. Hylton continued ranching and expanded beyond Mound Valley, buying other ranches including Rancho Grande, Devil's Gate, and the Reinhart ranches. He also ran two flour mills, one in Elko and the other at South Fork. But the Great Depression destroyed him. Having spread himself so thin, bankruptcy was his only option, and he sold all his holdings. He and his wife moved to Elko, where Hylton died in October 1947 at the age of ninety-three.

By the turn of the century, the town of Skelton consisted of a post office, a dance hall, a hotel, a restaurant, a store, a blacksmith shop, and a saloon. Sam Guldager, whose wife and daughters ran the town's restaurant, took care of the blacksmith shop. The Mound Valley Lodge of the International Order of Good Templars was located just west of town, and in 1905 the Templars hall moved into town. In 1900 the population of Skelton and Mound Valley stood at 109. In November 1902 a gunfight—a rarity for Elko County—took place between Fred Stone, a cowboy from the Carville Ranch, and Charles Conley, a saloonkeeper with a reputation for obnoxiousness. After a running argument, Stone challenged Conley to a fist fight. Conley came out with a gun, slugged Stone with it, and shot the hat off the unarmed Stone's head. Stone bought a gun for protection at the store, but when he prepared to mount his horse, Conley renewed the argument. Stone fired five times, hitting Conley four times. The shots killed Conley, who had just married local girl Hannah Guldager. The shooting was ruled as self-defense, and Stone was released.

Beginning in 1902 Albert Hankins, formerly a rancher in Ruby Valley, became Skelton's main businessman. In 1912 Hankins and his wife, Julia, bought everything in town except the Hylton store, which he purchased in 1916. In 1916 Hankins also built a new two-story brick hotel, which remains the town's most prominent landmark. In addition, he built a dance hall and erected large sheep corrals to capitalize on the sheep herds traveling through the area. With the arrival of prohibition in 1918, he closed his saloon, delivered the alcohol to federal agents in Elko, and converted the building into a small store. He also reestablished the post office on December 18 and renamed the town Jiggs, after the comic strip character. As a result, a local women's organization was named the Maggie Club, after Jiggs's wife. The original Hylton store burned on August 25, 1919, and was never rebuilt. When Hankins died in 1922 his holdings were sold, and Jiggs slowly faded as buildings burned or were torn down. The hotel became a residence and the school a post office.

In 1938 Jiggs's favorite son, Ted Carville, was elected governor of Nevada and served two terms. Later, he served as a U.S. Senator for two years. In 1940 the movie, *Brigham Young, Frontiersman*, which starred Tyrone Power, John Carradine, Mary Astor, and Vincent Price was filmed at Jiggs, and during the

1940s Emmett Cord, manufacturer of the Cord automobile, owned the Circle L Ranch.

A double murder took place at the Jiggs Bar in March 1962. Owner Andres Echevarrieta was found dead, and a local ranch hand, George Davis, was wounded so badly that he died the same day at the Elko hospital. The only clue to the killing was a saucer of milk on the bar floor. Bar patrons remembered a young woman who came into the bar with a bobtailed cat. At the same time, an Elko couple reported that they had been robbed by Lloyd and Mary Lancaster, who had been staying with them. Items taken included a bobtailed cat. Authorities arrested the couple in Weiser, Idaho, but charges against Mary were dropped when her husband admitted he had forced her to help. Lloyd Lancaster was sentenced to life without parole but was paroled in August 1978 despite many objections, including some from the panel of judges who had sentenced him.

Electricity finally arrived in Jiggs in 1963, courtesy of the Wells Rural Electric Company. That was also the year in which the town became nationally known, when all of its nine residents were loaded into a Volkswagen van as part of a promotional campaign. The post office, which still contains the original mailboxes, finally closed on December 15, 1975. A few people still call Jiggs home. The bar is open, and a small school operates. The community hall, a brick hotel, and a school/post office are among the remaining buildings. The two-story brick hotel, even after all these years, is still one of the most impressive buildings in Elko County.

Lamoille

(The Crossroads) (Walker Station)

DIRECTIONS: *From Elko, take Nevada 227 and follow it for 19.6 miles to Lamoille.*

Many people recognize the name Lamoille because of the popularity of Lamoille Canyon, but few visitors to the area are aware of the town of Lamoille's rich history. Two of the original incorporators of Austin, Thomas Waterman and John Walker, first settled the area in 1865. Waterman named the valley for his home in Lamoille County, Vermont. He built the first home in Clifton, located just below Austin, and after moving to Lamoille Valley served as constable for the Lamoille precinct. Until a jail opened in Elko, Walker had to bring prisoners to Austin. He died in 1897. Waterman left Lamoille in 1875 and died in 1881 in Tombstone, Arizona.

Other homesteaders soon followed Waterman and Walker, and in 1869 the town of Lamoille began to form after Walker built a store, a saloon, a blacksmith shop, and the Cottonwood Hotel. The complex, called Walker

Pete's Cabin, located at the head of the flume in Lamoille Canyon. (Noble and Roberta Skelton collection, Northeastern Nevada Museum)

Station, was located at what was known as the Crossroads, where the Fort Halleck Road met another road that ran down the valley. Some reports erroneously state that the California Trail intersected with the Crossroads, but in fact the trail did not come within ten miles of Lamoille.

The town's first election took place on June 21, 1869. Twenty-five voters participated, and as a result of the election, the Lamoille voting precinct organized on July 7. The town grew quickly and became the center of commerce and entertainment for residents along the west slope of the Ruby Mountains. A school opened in 1871 in one of the Walker and Waterman buildings, with Mary Trueman serving as teacher. A new school was built in the Read Field the following year, and children attended until 1898, when Charles Noble bought it and moved it to his ranch. By 1874 there were fifty students at the Lamoille school. Since only five students were required to establish a school, many valley ranches had their own. In 1886 the Humboldt school opened near the Pixley Ranch, later moved to the Voight Ranch, and remained in use there until the 1950s. The old school is now part of a house in Lamoille. The Valley View school operated at the Dysart Ranch from the early 1920s until 1939. The Pappas school was located on the Pappas Ranch for a number of years until it closed in the 1930s. The Patty Ryan school opened at the Ryan Ranch in 1914 and served students until 1920. Once the original Lamoille school moved in 1898, local residents purchased the William Wines store, which served as the school until a new one built in 1913 burned in 1919. Classes were then held in the Hankins-Bellinger Saloon until the Rabbit Creek and Humboldt schools moved to Lamoille in 1920. A new school, built in 1923, led to the return of the replacement facilities. The new school, which stands in Lamoille today, remained open until the Lamoille school district was abolished in 1961.

A post office, with Walker as postmaster, opened in Lamoille at John Walker's Crossroads Saloon on August 27, 1872, but closed on October 21, 1874, when Walker left Lamoille. Local citizens ran a volunteer post office until 1880, when Hank Keith bought the Crossroads Saloon and Cottonwood Hotel and reopened the office on May 10, 1880. However, when Keith sold out in 1882 the post office closed on August 2. It reopened for good on May 14, 1883, first at the Judson Dakin Ranch, and later in Lamoille itself. Peter Munson delivered the mail for $398 per year.

By 1880 Lamoille Valley's population was 207. Ralph Streeter, who ran the Lamoille, Pleasant Valley, and Elko stage line, served the town. In 1884 his brother, Sylvester, had Charles Gardner build him the first frame house in Lamoille. The Lamoille Cemetery was organized on land donated by Thomas Cahill, who said he would give the ground, since there was not much there a cow could eat, anyway. During Lamoille's period of growth, The Grove, a picnicking and recreational park, became very popular with Elko residents. Lamoille Valley developed into a prosperous ranching and farming area. One local farmer, Edwin O'Neill, achieved fame when his potatoes won first prize at the 1893 Chicago World's Fair. In 1899 J. B. Gheen opened a general merchandise store and reopened the Cottonwood Hotel, which had a picnic area and orchard.

By 1900 Lamoille's population had shrunk to 147, but after the turn of

Lamoille Creamery, 1908. (Elko County Library collection, Northeastern Nevada Museum)

the century people started a number of new businesses in the town. In 1903 Joe Lindsey built the popular Lamoille Mercantile Hall that also served as a dance hall. In 1907 the Lamoille Mercantile Company purchased the hall. Dances and home talent and traveling vaudeville shows took place in the 60- by 120-foot building, until it was torn down in the 1940s. Lamoille's first church, the Little Church of the Crossroads, was dedicated on November 5, 1905. While services had taken place at various locations throughout the valley since 1872, the Lamoille Presbyterian Church was not formally organized until October 1890. The congregation gradually became inactive and was not reorganized until 1904. Members raised $3,000 to pay for the beautiful church's construction. Its centerpiece was two memorial stained glass windows donated by the Greenfield family. The congregation quickly grew to fifty people, but by the late 1930s and early 1940s interest had faded, and regular services stopped. The church, while basically inactive, was not officially abandoned until 1954, but it had fallen into a state of sad disrepair. It was the site of some services during the late 1950s, but the deteriorating church was not the best setting for worship. However, starting in the mid-1960s a slow but steady restoration began under the guidance of the Lamoille Women's Club. Many county residents also donated time, money, and materials towards the revamping. A grant from the Max Fleischmann Foundation was a great help, and after over a decade of work the church, one of the most recognized landmarks in the county, was completely restored. It is still used today.

In 1906 Nevada Supreme Court justice George Talbot founded the Lamoille Mercantile Company, which operated until 1920. Born in 1859, Talbot was elected district attorney in 1884, named a district judge in 1890, and chosen for the Nevada Supreme Court in 1902. Other partners in the venture were Charles Noble and James Holland. By offering lower prices than shops in Elko did, the store enjoyed a prosperous business. Fred Voight ran it initially, and James Bellinger later took over. In October 1915 a robbery at the store, which also housed the post office, stirred up some excitement in the town. The thieves blew the safe open with nitroglycerin, muffling the noise by piling up sacks of potatoes. About $800 was stolen. The amount would have been larger, but all except three of the mail boxes were blown open and the contents destroyed. The robbers were never caught. After the store closed in 1920 the building, which a fire destroyed in the 1940s, served as a residence. The front section housed a saloon.

The Lamoille Mercantile Company financed the construction of the two-story Lamoille Hotel in 1907. O.T. Hill built it at a cost of $10,000. In 1912 the hotel was enlarged, and the new wing was identical to the original wing. The building contained twenty rooms, a bar, a dining room, a billiard hall, and a small dance hall. It remained open until a fire destroyed it on New Year's Day, 1963.

Construction began on the Lamoille Creamery in May 1906. Hill built the creamery, which cost him $2,000, though the total cost, including equipment, was almost $15,000 because of the high cost of specialized machinery. At its peak, the creamery produced 4,500 pounds of butter a month. Thirty-eight ranchers supplied milk, and the creamery paid $4,000 a month in revenues. After fifteen years local ranchers lost interest in milking, and the creamery closed in 1922. A. W. Sewell later used the building as a slaughterhouse. Some time afterwards the owners moved it to the Horseshoe Ranch at Beowawe and converted it into a meat house.

In 1913 Charles Noble built the Lamoille Roller Mills and hired W. L. Barclay as miller. While the mill was initially successful, war rationing hurt the business, and it never completely recovered. By 1924 the mill was the only active one in Elko County. During the 1920s local ranchers gradually shifted from growing grain to producing hay, and the mill closed for the first time in 1927. After that, it only operated for a short while in 1928 and 1936 before being dismantled. Part of the building was converted into a home.

A unique element of Elko County's history is the flume and power plant built in Lamoille Canyon in 1912. The construction plan, who ran the W. T. Smith Company in Elko and brought the first telephone lines to Lamoille in 1898. He surveyed the canyon in 1903 and sold the plan and survey to a group of Elko businessmen who formed the Elko-Lamoille Power Company. The company strung power lines to Elko in 1913 and to Lamoille in 1914. The wood flume down the canyon was 15,000 feet long and started at a place called Pete's Cabin, home of Pete Arcimes, the water master who controlled the flow. The flume had to be rebuilt in 1944 and replaced with steel pipes in 1956. The Lamoille Canyon plant served the town for many years, but demand outran supply because during the summer months there was insufficient water to produce enough electricity, and additional power had to be purchased elsewhere. After the power plant burned in 1971 it was not rebuilt. Because of U.S. Forest Service regulations, the flume, Pete's Cabin, and all other buildings were removed. Little remains today to show the importance of this vanished enterprise.

Over the years, Lamoille has remained a tightknit ranching community, though old buildings have disappeared and new ones have been built. In 1940 Robert "Doby Doc" Caudill purchased the old log Cottonwood Hotel and moved it to his Frontier Village in Las Vegas, from where it long ago disappeared. One of the buildings from the Tonopah Air Base came to Lamoille after World War II and still serves the Lamoille Women's Club, which hosts the annual Lamoille Arts and Crafts Fair—a major local attraction in June. The completion of the paved road from Elko to Lamoille in 1947 greatly increased Lamoille's accessibility. Lamoille and Lamoille Canyon continue to draw both local residents and tourists from all over the country. The area is one of the most beautiful in Nevada. Lamoille remains much as it was a hundred years ago. Do not be surprised to see deer grazing in The Grove in the middle of town.

Larrabee Mining District

DIRECTIONS: *Located 6 miles east of Mineral Hill (in Eureka County).*

As a result of mining activity at nearby Mineral Hill, a number of prospectors explored the adjoining area. In the early 1880s many claims became active, and a mining district, Larrabee, was established. The district was named for Stephen and Albert Larrabee, brothers who had mines in the area. The first mines worked in the Larrabee District were the Argent, the Carbonate, and the May Gillis, all of which Frank Chandler owned. The district was most active in 1882, when more than thirty mines were being worked. They included the Montezuma, the Argentine, the Cortez, the Grace Darling, the Triumph, the Sacramento, the Mable May, the Dundee, the Argent, the Cornwell, the Liberty, the Clara, and the Nettie. There were as many as fifty miners in the district during the summer, but despite all the activity, hardly any rich ore was found, and by the following year everyone was gone. One more try was made in the Larrabee District in 1887, when Thomas Whelan worked the Woodside mine, but he had little success.

No more mining took place in Larrabee until the 1970s, when the Jay mine produced about 1,000 tons of barite. Some exploration for microscopic gold has occurred in the area, but no ore bodies were established. Only small ore dumps from the 1880s mark the site.

Lee

DIRECTIONS: *From Jiggs, head north on Nevada 228 for 8 miles. Exit right and follow the road for 4 miles to Lee.*

John Martin, a native of Maine who settled in the South Fork Valley in 1869, named Lee for Robert E. Lee. A real community was slow to develop there, although a number of ranches were homesteaded during the next decade. The farmers and ranchers found ready markets in Elko, Tuscarora, and Eureka. The town of Lee began to form when the Eureka Flour Mill Company was organized in March 1881. Under the guidance of John Ainley, president, and superintendent John Martin, a conglomerate of merchants and farmers raised funds to begin construction of the South Fork Flour Mill in May. The three-story building was 40 by 60 feet and also contained a 30-by 100-foot grain house. Water from the south fork of the Humboldt River turned a fifteen-foot turbine water wheel to crush the grain. The mill's total cost was $19,000. It began operation in November, producing eighty barrels of flour a day.

As a result of the mill's construction, the small town of Lee soon had a population of about fifty, and a school was opened. A dugout on the side of

Lee Store, 1960. (Tony and Ellen Primeaux collection, Northeastern Nevada Museum)

a hill, with a front end made of sod, it contained only benches and no desks. However, the influx of people into Lee necessitated the construction of a new school near the mill. A post office opened on February 14, 1882, with Martin, the mill superintendent, as postmaster.

After a few years, three local farmers by the names Henry Jones, Edwin O'Dell, and Fred Drown bought the mill. The trio ran it until 1898, when they sold out to W. O. Williams, former owner of the Ruby Valley Flour Mill. Williams produced two brands of flour, Lily White and Blue Ribbon. He retired in 1910 and sold the business to his sons. One of the sons, Will, was later killed by the large flywheel at the mill. Not long afterwards, the mill caught fire and burned to the ground. J. J. Hylton of Jiggs rebuilt it, and Lee's residents used the roomy building as a dance hall. In 1915 Hylton sold the mill to the Lowell Grain and Milling Company, but, unable to make a go of it, the company sold out in September 1917 to the newly formed Elko Milling Company, whose directors included W. W. Percival, J. J. Hylton, John Henderson, C. B. Henderson, and Robert Hesson. Because of World War I and the resultant war rationing, the mill was no longer useful, however, and it soon closed. The mill, used as a recreation hall after its closure, remained standing until the 1930s, when it burned.

Another venture in Lee, though a short-lived one, was the South Fork

Creamery. It opened in 1910 but closed a couple of years later when the cows that its owner, Fred Drown, had used contracted tuberculosis and had to be destroyed. After the closure, the building housed a store run by Ed Williams, who also ran the post office from his business.

Once the flour mill had shut down, the small town of Lee survived as a social center for South Fork Valley families, but its population shrank over the years. Improvements arrived, though, in the form of phone lines strung in 1911, and electricity brought in during 1963. The school district, however, was abolished in 1957. In 1935 Lee and the South Fork Valley began a slow transformation into the present-day Temoke Tribe of the Shoshone Reservation. Very few ranches independent of the reservation still exist in the valley.

While ranching and farming have long been mainstays of Lee's local economy, mining has made its presence felt since 1869, when the American Beauty mine began operations. After producing copper valued at $1,200, however, it closed in 1872. Little production took place in the area until 1908, when the Delmas Copper Company found paying ore while digging a 170-foot tunnel in its mine. This was the beginning of an extended period of activity during which mines produced values in copper, lead, and silver every year from 1915 to 1929. As a result, the South Fork Mining District was organized in 1918. The Onondago Mines Corporation, based in New York, was the main company at the time, and its property produced good quantities of silver and lead. Other active properties included the American Beauty mine, the Hargrove mine, the Skagg claims, the Sabala claims, and the Nick Jayo group.

The 1920s was the district's most productive era. In 1923 both the American Beauty and the Long Canyon mines were active, and $18,000 was spent on a road to the mines. In 1924 the C & J Mining Company purchased the American Beauty mine for $125,000, but after it had made two payments and invested $30,000, the company gave up and returned the property to Getchell and Associates, the original owners. In August 1924 a twenty-five-ton concentration mill started at the Long Canyon mine. G.C. Hargrove, owner of the Hargrove mine, and Bob Skaggs, owner of some adjacent claims, had been embroiled in a long-standing feud that came to a head in 1925. Skaggs had claimed that the Hargrove mine was his, but the courts decided in favor of Hargrove in June. Still believing the mine was his, Skaggs had Hargrove arrested for threatening him, but investigators believed it was just a ploy on Skaggs's part to try and wrest control of the mine from Hargrove. Skaggs had Hargrove arrested again in July. That was enough for Hargrove. He optioned the mine to E.G. Spaulding, who later purchased the property for $25,000.

In July 1927 the American Beauty Mining Company began construction of a concentrator that it completed in October, but by the end of 1929 all the mines in the district had shut down. In 1949 the Knob Hill mine (in Long Canyon) was developed, and it produced until 1958. This was the only mine that produced in Lee since the end of the revival of the 1920s. While some exploration has occurred there since 1958, there has been no production. Total ore production includes 1.8 million pounds of lead, 123,000 pounds of zinc, 24,000 pounds of copper, and 11,000 ounces of silver.

Lee is a quiet town with a population of about fifty. A number of old buildings, including the school and creamery, remain. The Oregon Shortline Railroad's Contact depot stands at Lee and is the only depot of the line left in Nevada.

Mud Springs
(Medicine Springs) (Dead Horse)

DIRECTIONS: *From Harrison Pass, head north on Ruby Valley Road for 1 mile. Take a right and follow the road for 12½ miles. Turn right again and follow that road for 6 miles. Turn right for the Golden Pipe mine. Continue on the main road for 1 more mile and turn right. Follow the road for 2 miles to the Mud Springs mines, which are scattered along the road for another 2 miles.*

Sam Backman, Garfield Bardness, and Fred Martin made the first discoveries in the Mud Springs District in 1910. The Nevada Dividend Mining Company discovered the district's biggest producer, the Dead Horse mine, in January 1911. It later became known as the Silver Butte mine. In October

One of a number of buildings scattered around the Mud Springs mining district. (Photo by Shawn Hall)

A.D. Thirot, Sam Backman, and Fred Martin began working a new group of claims that would eventually become the Silver King mine. Lorin Morrison bought the Dead Horse mine and other claims in February 1913. While a central camp did not develop, by the beginning of 1914 there were fifteen tent houses at Mud Springs and thirty men scattered throughout the district. But very little ore was actually shipped. Only a small load, sent to Salt Lake City in 1915, was shipped during the teens. One of Mud Springs's problems was that the nearest railroad shipping point was Currie, which was almost forty miles away.

By 1920 three companies were active in Mud Springs. They were the Nevada Dividend Mining Company, the Butte Valley Mining Company, and the Nevada Garfield Mining Company. This was Mud Springs's first real year of production, though considering the amount of activity that appeared to be taking place, production was always limited. In 1923 the Monitor mine produced twenty-five tons of ore, and the McMullen claim was also shipping ore. The Silver Butte Consolidated Mining Company, which Morrison represented in Mud Springs, was also organized in 1923 with Moses Taylor as president and D.A. Walton as manager. The company immediately built a road to Currie and produced 4,201 pounds of lead and 768 ounces of silver.

In August 1924 Morrison and a partner, Milton Odgers, formed the Mud Springs Mining Company. Walton was also this company's superintendent, while Morrison had earlier been instrumental in developing the Polar Star mine at Warm Creek. A hoist and air compressor were installed at the Silver

Mill ruins at Medicine Springs, Mud Springs mining district. (Photo by Shawn Hall)

Headframe of the Golden Pipe Mine, complete with eagle nest near top guide wheel. (Photo by Shawn Hall)

Butte mine, which employed eight people. The claims of a new company, the Marie Consolidated, produced $600 worth of lead and ore.

By 1925 only the Silver Butte Company and Fasano claim group were producing. The best of these early years were 1926 and 1927, during which the mines produced 131,000 pounds of lead and almost 2,000 ounces of silver. In May 1929 the Silver Butte Company completed a 100-ton concentration mill that used oil flotation. The first 600 tons treated generated $2,250 in revenue, but the mill did not treat much more before the company closed it down and dismantled it. In July Jessie Sharp and Robert Morrison sold a group of claims to P. H. Kelley, H. C. Purdy, and E. J. Nelson for $60,000, but the new owners found little of value on their claims. Hope for the district revived when the railroad listed Mud Springs as a potential stop on the planned Spruce Mountain to Ventosa railroad spur, but the line never materialized. The mill's failure forced the Silver Butte Company to fold, and by 1930 the district was empty except for a prospector or two.

Not much else happened in Mud Springs until September 1934, when Robert Morrison and Barney Horn sold twenty-four mining claims to William Bodfish for $30,000 and a royalty. But Bodfish defaulted in the spring, and Morrison and Horn resold the property to the Clark brothers for $200,000 in July 1935. The brothers were soon shipping fifteen tons of ore per week. In March 1936 Barney Horn, who was still working the district, was arrested and charged with the theft of forty-seven dollars worth of silver-lead ore from Verl Dastrup—who was working the Silver Butte mine. Horn was acquitted after only thirty minutes. It turned out that a miner named Martin Smith, who Horn had grubstaked earlier in the year, had set him up. Horn, who had previously mined in Tuscarora and Aura, left the district and purchased the Deluxe Hotel in Elko, but he died that September. In 1937 the Silver Crown group's mines produced $2,500 worth of silver and lead. This represented the district's last bout of real production until 1950, though some mining did take place there in the 1940s.

In 1950 a new Silver Butte Consolidated Mining Company organized, and a sixty-ton concentration mill began operations at Medicine Springs, six miles to the west. The company produced $10,000 in 1950, the district's best year, but it closed down operations in August 1950. Two security guards were the only remaining employees. Little was mined during the next couple of years, and since 1956 the area's mines have produced nothing. The value of the district's total production stands at less than $45,000. While the area was drilled and explored extensively in the 1980s, no new mining has taken place. The district is enjoyable to visit. At Mud Springs many miners' cabins, gallows frames, mines, and other buildings are scattered over a wide area. Be careful, because there are some unmarked shafts that could be dangerous. The mill's extensive foundations mark the site at Medicine Springs.

Racine Mining District

(Superior Mining District)

DIRECTIONS: *Located on the White Pine–Elko County line near the Jiggs Road.*

Very little is known about the Racine Mining District. No mining books mention it, but the *Elko Independent* gave it a great deal of coverage. When the author located the site, it was difficult to determine whether it was in Elko County or White Pine County. It is included in this book because of its proximity to Elko County.

The district first became active in the spring of 1872. It was named Racine after the town in Wisconsin from which Albert Dewitt, one of the area's miners, came. In the summer the George Washington, the Uncle Sam, and the Star of the East mines began operating. On August 17 the district was renamed Superior. George Henderson was president and his brother, R.M., was recorder. The Cambrian Gold and Silver Mining Company was organized and took over all the mines in the district. In August 1873 the district expanded, and in July 1874 ore sent to the Newark mill assayed as high as $600 per ton. But by 1875 the district had been abandoned. In 1881 James Crawford, William DeGraff, John Mayhugh, and George Grayson made a last try when they worked the Stella mine, but the mine ran out of paying ore.

It took many days of exploring before the author finally located the Superior District. It is apparent that the limited mining that took place here ended in the 1880s. Only small ore dumps mark the site, and there is no evidence of further mining attempts.

Ruby City

(Lurline) (Fairplay)

DIRECTIONS: *Located 7 miles north of Arthur.*

While the promotional town of Ruby City did not come into being until 1913, the area is historically important in another way. On February 7, 1879, the Fairplay post office, with Marshall Lemmon as postmaster, opened in the area at the ranch of C. W. Grover Jr. The ranch was also a stop on the Ruby Valley stage, which Kelly and McCain ran. The office remained open until August 10, 1893.

Ruby City, a planned community of seventy-five families, began to develop in 1912 when a group of Utah land promoters purchased 5,000 acres, the bulk of which was the George Robinson Ranch. The group subdivided the land into 40- to 160-acre lots and built a five-mile gravel road named Taylor

Avenue, after one of the promoters, leading to the townsite. The promoter slyly named the town's streets for local ranchers in the hope that they would relocate to the town, but it never happened. Within a year, seventy tent homes had been erected at Ruby City. In 1914 a ten-inch-wide, five-mile-long canal was built to bring water to Ruby City, both for the purpose of irrigation and for the townspeople, but it was never used. Dry years prevented water from reaching the town, but a large hotel opened its doors nevertheless. The original promoter could not keep up payments on the property, and the Glenn brothers—John, Tom, Heber and Walter—took it over.

By 1915 the town contained a hotel, two schools, a store, a blacksmith shop, and a Mormon church. A post office, named Lurline in honor of post-master John Glenn's sister, opened on May 28. A covered wagon, making numerous trips, brought more than fifty students to the schools. Edith Wyman and Lewis Sharp served as teachers. But Ruby City was doomed from the start. Without adequate water, the planned, but dry, farming community could not develop. In addition, there was too much alkali in the soil, and the growing season was too short. By 1918 almost all the residents had left. The post office closed on September 30, 1919, and the schools followed suit in 1920. By then Ruby City was already a ghost town. In 1921 Edwin Goodwin, who had loaned the Glenn brothers money for the town's purchase in 1914, reclaimed the property, which the Neff Ranch still uses as range land. Not much remains at Ruby City, because very few permanent structures besides the hotel were ever built there. All those constructions that were built have been moved or torn down. Scattered debris, foundations, and the faintly visible roads are all that is left.

Sherman

(Walther's)

DIRECTIONS: *From Jiggs, continue south for 16½ miles. Turn left and follow the road for 6 miles to Sherman.*

Valentine Walther and his wife, Sophie, homesteaded along Sherman Creek in January 1876. They had come from Diamond Valley, in Eureka County. Because of the harsh conditions, the couple and their four children had to live in covered wagons until they completed a home in the summer. The Walthers had twelve children; they were Sophie, Annie, Louise, Lizzie, Valentine (who died at two months), Joe, Susie, Mary, Dora, George, Fred, and Cora. Because he had so many children, Valentine established the Sherman School District no. 31 on May 9, 1886. A log cabin on the ranch housed the school. For many years, Sherman was a stop on the stage line heading south to Hamilton, and the Walthers provided meals and overnight lodging.

Walther had one of Nevada's best orchards. He raised cherries, plums, apricots, and apples and at one time had more than 250 apple trees. He also maintained a large vegetable garden and grew raspberries, strawberries, and grapes.

Sophie met with a terrible accident in June 1895, when she and Valentine were on their way to visit neighbors. Valentine got out of the wagon to open a gate. When the horses pulled the wagon through, a breaking branch startled them, and they took off. Valentine chased the team and found the overturned wagon half a mile down the road. The wagon tongue had struck the ground, flipping the wagon and hurling Sophie more than twenty feet. She did not survive the night, and Valentine was left to raise eleven children alone.

In 1902 Walther and Nick Scott began building a huge, two-story home of eighteen-inch logs. The nine-room home took almost two years to complete. The second floor featured a large hall that was used for dances and other gatherings. The structure was and still is one of the most impressive ever built in Elko County. A post office, called Sherman, opened on February 28, 1903, with Valentine as postmaster. While the Sherman school was still opened, a new school, the Dewey, complete with red brick siding, opened below Sherman near the edge of the valley. The Dewey school closed in 1915. The post office closed on June 15, 1915. Walther sold the ranch in 1922 and retired to Elko, where he died in 1933. At the time, he had thirty-eight grandchildren and eighteen great-grandchildren. The Sherman school, which operated for fifty years, remained open into the 1930s.

Owners of Sherman since Walther sold the property have included the Peters brothers, George Bray, George Fairchild, Doug and Barbara Mitchell, and Peter Scheidermann. The ranch's current owners now raise buffalo. While a new house has been built, Valentine Walther's two-story log home still dominates Sherman. The Elko Chamber of Commerce hopes to preserve the building by moving it to Elko, where it would serve as the chamber's offices. If sufficient grant money is raised, the plan could become a reality. The original log schoolhouse still stands as do a few other buildings. Sherman is on private property, behind a gate. Please do not enter without permission.

South Fork
(Shepherd's Station) (Coral Hill)

DIRECTIONS: *From Elko, head east on Nevada 227 for 6.9 miles. Head south on Nevada 228 for 11 miles. Turn right and continue ½ mile to South Fork.*

George Crane, John Richardson, William Tucker, Thomas Chandler, and Robert Toller first settled the South Fork area in 1867. They made two trips to Austin each year to stock up on supplies. In the fall of 1868 George Shepherd organized the Denver–Shepherd toll road, also known as the Hamilton road, which ran to Hamilton. It was the first toll road to be built in eastern Nevada. At the time, it was the only road heading south from

Elko to Hamilton, but the Gilson turnpike, which was completed on the east side of the valley the following spring, ended Shepherd's monopoly. Until Elko formed in 1869, the Taylor and Morton stage used Shepherd's Station as its terminus. Taylor was well-known for his ability with a shotgun and was so feared that no one ever harassed his stages. Shepherd himself occasionally road shotgun for Wells Fargo. He settled at South Fork, and his station was the first overnight stop south of Elko.

In 1869 Shepherd built a bigger stage station and a hotel that contained a bar, a dining room, and a large wine cellar featuring French wines. A post office named Coral Hill opened at South Fork on March 25, 1870, but closed on July 19, 1871. Shepherd's toll road fell victim to the vicious stage wars that took place during the 1870s. When Hill Beachey took over the Gilson road, his lower prices and faster route attracted business away from Shepherd's road. In 1870 when the Woodruff and Ennor line changed to run on the Gilson road, virtually everyone stopped using Shepherd's route. Not one to give up, Shepherd opened a new toll road leading from Elko to Bullion in 1870. On February 27, 1874, the post office reopened with the name South Fork. On May 1 it reverted to its old name, Coral Hill. The post office remained in operation until March 14, 1877, and Shepherd was postmaster during this entire period. After some time had passed, Shepherd quit the stage business and returned to ranching, but his hotel stayed open and was the "in" place for Elko residents to go for dances, parties, holiday gatherings, and weddings. Shepherd always promised that the table would be full and the bar stocked. The county purchased Shepherd's old toll road in 1882 and

The William Crane Ranch, 1886. (Reginald Heard collection, Northeastern Nevada Museum)

made it into a public road. Shepherd later served as Elko County Treasurer and served two terms as state senator.

In 1875 Shepherd discovered the Mineral Soap mine. The Elko Mineral Soap Company, which marketed the soap, shipped it to a soap factory in Oakland, California. The Shoshone word *san-too-gah-choi,* meaning "good soap," was written on the wrapper. The soap won an award at the 1894 World's Fair in Chicago, since it was the only product on display that had come from a natural soap mine. The soap stayed on the market until 1900.

All of Elko County mourned when George Shepherd died in 1885. While memories of Shepherd's Station have faded over the years, many ranches still call the area home. With the completion of the South Fork Dam, the area is now a popular recreation spot. Not much remains at Shepherd's Station except for some foundations and scattered debris.

Valley View
(Beryl City) (Hankins) (Dawley Canyon) (Gilbert Canyon)

DIRECTIONS: *Located 5 miles north of Harrison Pass.*

John Hankins, a Lamoille Valley rancher, made the first discoveries in the Valley View District in 1913. A small camp called Hankins formed during the summer, but the bismuth and scheelite in the area were not valuable enough to justify continued mining. Atkins, Knoll, and Company of San Francisco took over the mine in 1916, but they, too, did little. In 1925 the Mutual Mica Company, which Roy Rigsby owned, began extensive mica operations, shipping an average of 500 pounds a day. Fred Bender discovered the mines in 1921, but by the summer of 1927 they had run out of good mica, and the Mutual Mica Company folded. C. W. Kennedy bought the property at a sheriff's sale in October for $2,900, but all in all the district inspired little interest until 1931.

In April 1931 M. W. Young and A. R. Cayton found some strange crystals on the old Mutual Mica mine dumps. They turned out to be a beryl deposit. The property was optioned to the Columbia Mining Company of Pittsburgh, Pennsylvania, and a new mining camp, Beryl City, sprang up during the summer. An air of great excitement surrounded the camp, which a constant flow of curious people visited. In 1932 the operation consisted of thirty-three claims. The Gilbert Canyon Mining District formed, but beryl was not a very valuable commodity, and the camp and mine was abandoned by the following summer. Mining activity returned to the district in the 1940s, when the Valley View mine began tungsten production and produced 150 units in 1943. In March 1943 Young, who was still working the district, organized

Beryllium of Nevada to further explore his beryl deposit. He and his partners, A.J. Errington and A.E. and R.H. Downer, developed the deposit, which now covered 108 claims. But samples sent to government labs once again confirmed that the beryl was of poor quality. The property was leased in February 1954 for $500,000, but it produced only a small amount of tungsten. In 1956 Vincent Gianella, prominent Nevada geologist, checked out the beryl property, but he, too, arrived at the conclusion that the ore quality was very poor. Little production ever took place in the district, though interest in it continued. Nothing much has happened there since the 1950s. Mine dumps and remnants of a few cabins are left in the Valley View District.

Warm Creek
(Haws) (Polar Star)

DIRECTIONS: From Wells, head south on U.S. 93 for 24 miles to Warm Creek Ranch. The mines are 2 miles to the southwest.

Rancher Albert Haws established the Warm Creek Ranch in 1858. Haws was an unsavory character. People believed that he had a special bond with local Indians, and that he encouraged and helped them to attack and rob immigrants traveling on the California Trail. Eventually, Haws got his comeuppance. In August 1869, with no provocation, he murdered another Clover Valley rancher, Thomas Dunn. Haws then fled to Utah, where he killed a Utah deputy U.S. marshal and wounded Elko County Deputy Sheriff J.F. Corrigan. A Tooele County posse finally cornered him, shot him seven times, and killed him. Shortly afterwards, his ranch was sold for $26.79 to satisfy back taxes. Horace Agee made the ranch part of his empire in 1901, when he purchased it from C.H. Ballard. Agee and his wife, Etta, also operated a roadhouse for Ely-bound travelers and offered meals for fifty cents. A post office operated at the ranch from October 24, 1907, to November 15, 1909. The postmaster was Flora Bebee.

Mining became big news at Warm Creek in 1912 when T.J. Franks, M.E. Reed, and B. Woolverton discovered the Polar Star mine in March. The Bluedale, the Tribrook, the Mammoth, and the Tiger mines had limited production in the mid-1880s. In June L.N. Morrison, owner of many claims in the Mud Springs Mining District, bought the property and organized the Polar Star Mining Company. He hired ten men to develop and work the mine. They found some silver and lead, but the mine's major mineral was zinc. In December 1915 the Clover Mining Company formed to work some nearby claims, which later became the Echo mine. Also in 1915 Morrison sold the Polar Star mine to the Nevada Zinc Mining Company but stayed on as general manager. The mine employed twenty-five men, and the miners even formed a baseball team. In July 1916 Morrison resigned as general manager, and B. F. Grant replaced him. Soon afterwards, mining operations were cut back dramatically, but limited work continued until 1918 when the company folded. From 1915 to 1918 the mine produced 1.2 million pounds of zinc, a far cry from Granger's book, which reports a low and erroneous estimate of 100,000 pounds.

After 1918 the only attempts made to work the Polar Star mine were tentative ones in the 1930s and 1940s. From 1941 to 1943 Charles Stoddard and Jay Turner reworked the Polar Star dumps and shipped the ore through Tobar. In 1953 Hadel Products Company built a small mill to treat old ore, but it was unsuccessful and shut down after producing a very small amount. Some exploration has taken place at Warm Creek since then, but as of 1995 there has been no more mining activity. The area's total production stands at 1.8 million pounds of zinc, 41,000 pounds of lead, and some silver and copper.

The Warm Creek Ranch continued to be an important element of Clover Valley. The Agees sold it to the Badt family in 1915 after purchasing the Wiseman Ranch to the north in Clover Valley. Edward Lees and three Badts— J. Selby, Milton, and Herbert—formed the Warm Creek Land and Livestock Company, which became known for its thoroughbred shorthorn cattle. A school, with Zelma Berry as teacher, opened at the ranch in 1918. The large ranch house, which also housed the school, caught fire and burned to the ground in April 1922. J. Selby Badt and an employee, M. B. Goldstone, tried to put out the fire, but could not. In October 1929 Samuel and David Neff and C. S. Woodward, friends of the Badts who were involved in similar mining interests at Spruce Mountain, bought the ranch. After the sale took place, the new owners always welcomed the Badts at the ranch, and they spent many vacations there. In fact, it was at Warm Creek that Herbert Badt died in February 1932, after a long illness. After being open off and on, the school closed in 1934, when only five students were left.

Subsequent owners of the Warm Creek Ranch have included Pete Wollman, Mark Scott, and Vernon Westwood. Scott paid Wollman $95,000 for the ranch in August 1953, but Wollman died only days after the deal was made. There was a CCC camp at Warm Creek from 1937 until it was disbanded in November 1941. Following the closure, rumor had it that the old camp would become a Japanese internment camp, but that never happened. The ranch still operates today, and it includes a couple of older buildings. A collapsed shaft, mine dumps, and some small buildings, which probably date back to the 1940s, are left at the Polar Star mine.

Southeastern Elko County

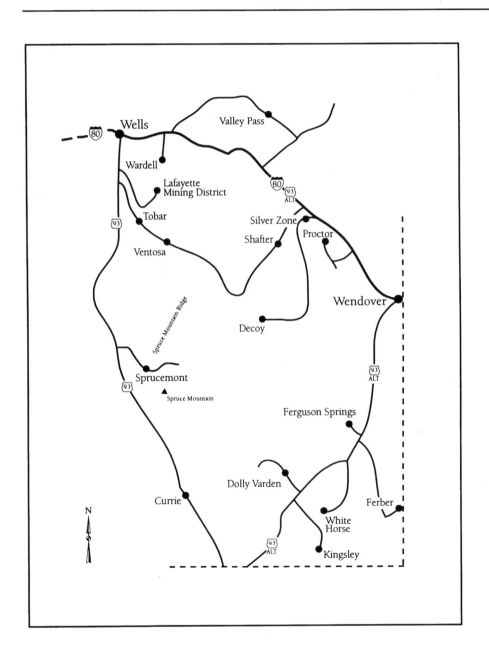

Wells
Valley Pass
Wardell
Lafayette
Mining District
Tobar
Silver Zone
Shafter
Proctor
Ventosa
Spruce Mountain Ridge
Wendover
Decoy
Sprucemont
Spruce Mountain
Ferguson Springs
Dolly Varden
Currie
Ferber
White
Horse
Kingsley

N

*Nevada Northern
Railway train pulling
into Currie, 1950. (Anita
Cory collection,
Northeastern Nevada
Museum)*

Currie

(Bellinger's Spring)

DIRECTIONS: *Located 62 miles south of Wells on U.S. 93.*

The springs at Currie, which were being used long before the town
formed in 1906, were originally a stop on the Toano and Cherry Creek road.
The town of Currie formed in 1906 when the Nevada Northern Railway was
constructed. The rails arrived in Currie on May 18, and the first passenger
train came on May 22, though the railroad was not completed to Ely until
September. Tri-weekly service started on June 2, and the railway built a depot
that served passengers, shipped freight and livestock, and was the home of
the railway agent. The small town was named for Joseph Currie, who had
been the owner of a ranch on nearby Nelson Creek since the 1880s. The first
substantial building constructed was the two-story Currie Hotel, which Currie owned. A post office, which the hotel housed, opened on August 8 with
Currie as postmaster. Henry Phalan, a local rancher, soon built the Steptoe,
another hotel.

Currie quickly developed into the transportation and livestock center of
southeastern Elko County and maintained a consistent population of twenty.
The entire station was under quarantine for awhile when a diphtheria epidemic struck in July 1907, but prompt action prevented any deaths from
occurring. In 1908 Earl and Leona Reynolds moved to Currie to open a new
telegraph office and railroad agency station. They also opened a store with
Jim Byron. A one-room schoolhouse went up near the Phalan Ranch in 1908.

Two years later, because of increased demand, two more rooms were added. The teacher lived in one of them. The school also served as the Currie civic center and was the venue for local dances.

By 1910 Currie was in its heyday. Besides the depot, the hotels, and the school already in the town, a store and a saloon opened. In addition, the railway constructed a section house and turntable. Currie had a reputation as a haven for jackrabbits, and the Nevada Northern ran a special Sunday train from Ely to give about 100 people an outing of recreational hunting and rattlesnake drives. The town's namesake, Joseph Currie, who had lived in the area since the 1880s, died in October 1912 and was buried in Cherry Creek.

The Reynolds soon bought out Byron and resigned from the railroad to build a new store. The post office moved to the store, and Leona Reynolds became assistant postmaster to her husband. The Steptoe Hotel came down in the late teens, and the Currie Merchandise Store, which housed the post office, opened on the site. In the ensuing years, Currie continued to serve the railroad. When the road from Wells to Ely became a paved highway, a gas station and store were built.

In 1941 the Nevada Northern Railway discontinued passenger and mail service at Currie because of slumping business and also because the depot agent, Jerry Cormer, left town. The limited amount of business did not warrant finding a replacement. Freight and ore trains continued to rumble through Currie until the 1980s, while the post office stayed open until May 28, 1971. The original Currie depot still stands, as does the Currie Hotel.

Travelers in front of the Currie Hotel, 1920s. (Anita Cory collection, Northeastern Nevada Museum)

Some other buildings from Currie's early years have also survived. A gas station and store are in operation, and Currie has a highway maintenance station. The town's population is still about twenty.

Decoy

DIRECTIONS: From Shafter, head south along the Nevada Northern rails for 12 miles to Decoy.

Decoy came into being with the construction of the Nevada Northern Railway in 1906. It was only a siding until the Decoy Mining District was organized and began producing manganese ore from mines located eight miles east of the town. Loading platforms allowed for the loading of ore.

Before mining began there, the area around Decoy was a dry farming site. Dry farming—a type of farming that relies solely on rainfall and not on irrigation—achieved widespread popularity during the early and mid-teens. By 1915 as many as twenty dry farms were located around Decoy, but a combination of drought and rabbits caused the failure of 70 percent of the crops. While more established dry farming communities such as Metropolis and Tobar kept going, the relatively new farms at Decoy had a high rate of attrition. The following year proved even more devastating. Of the seven farms left, five completely failed. By 1917 all dry farming around Decoy had ceased.

In 1917 the Darky (Black Rock) mine, which Holmquist and Johnson owned, began producing manganese. In 1917 and 1918 it shipped 4,500 tons of ore through the Decoy siding, but that was the only significant thing the district produced. Duval International (now known as Battle Mountain Gold Company) did some exploratory drilling in the 1980s. They outlined a disseminated gold deposit, but it has not been mined yet.

Only ruins of the loading ramps and section house mark the Decoy siding. A number of old day-farming homesteads dot the area west of the railroad tracks, while at the Darky mine collapsed shafts, an open pit, and mine dumps remain.

Dolly Varden

(Mizpah) (Victoria) (Moores Camp) (Last Chance)
(Granite Mountain) (Watson Spring)

DIRECTIONS: *From Wendover, head south on U.S. 93A for 38 miles. Exit right and follow the road for 4 miles to Dolly Varden. Continue 8 miles to Victoria.*

Dolly Varden, a housing place for miners, was a small camp located about eight miles east of the mines at Mizpah (otherwise known as Victoria), part of the Dolly Varden Mining District. Silver was first discovered in the area in 1869, but no one did much about it. In June 1872 Johnson and Murphy discovered the Dolly Varden mine. The name Dolly Varden has a number of possible origins. At the time of the mine's discovery, the Dolly Varden Party, which was against political rings and bossism, was active in Nevada. Dolly Varden is also a character in Charles Dickens' book, *Barnaby Rudge*. In addition, Dolly Varden is the name of a red-spotted trout native to nearby Nelson Creek.

The summer of 1872 saw extensive exploration taking place at Dolly Varden. Two camps formed, one at Dolly Varden Spring (Dolly Varden) and another at Watson Spring, named Moores Camp, for J.L. Moore, owner of the First Chance mine. The two had a population of fifty by the end of 1872. Other active mines included the Anchor, the Dolly Varden, the Centennial, the William Penn, the Shoshone Queen, the Bronco, the Keystone, the

Jackson, and the Emerald. However, the best mine was the Victoria, which produced copper intermittently for more than 100 years.

A smelting furnace constructed at Dolly Varden was used for the next two years, but it was not successful and ore was later shipped by wagon. While production was limited, it kept the two camps growing. By 1880 the population was eighty-seven, and a school, with Bessie Buddy as teacher, opened. In July 1881 the Kinsley Copper Mining Company bought the Victoria mine. The mine, under the guidance of foreman Johnny May, developed into one of the richest copper mines in southern Elko County. Its ore was shipped to Baltimore, Maryland, and it employed twenty-five men. The Dolly Varden Mines Company (J. P. Murphy, superintendent and district recorder) formed in May 1883 and employed eighteen. The company began refurbishing the old smelter, but it caught fire and burned to the ground while the work was being done. John Phalan, who ran Phalan Station on the Cherry Creek road near present-day Currie, built a large boardinghouse at Dolly Varden. Some other mines that opened on Granite Mountain included the Atlanta, the Summit, the Black Prince, and the Surprise, but a severe drop in copper prices forced the closure of the mines at Dolly Varden. The district emptied and did not revive until 1905. F. J. Barrigan made new discoveries at Mizpah Spring, just north of the Victoria mine. New mines opened, old mines revived, and by June 1906, 135 new claims had been filed.

Sam Newhouse, millionaire copper baron, visited the mines and considered investing but never did. The most prominent of the new mines was the Morning Glory, which J. L. Moore, Charles Karbenstein, and Thomas Keough owned. The R. T. Root Company, which a Denver millionaire owned, bought the Victoria mine. A big copper strike was his immediate reward. He soon purchased mining properties owned by John Phalan, Warren White, J. Byron, F. Herrell, and Moore and McAuley. A tent town formed at Mizpah, and in January 1907 the Mizpah Mining District was organized. Herrick and Company platted a townsite on the Barrigan and Brady property, and the company sold $22,000 lots in one day. With the completion of the Nevada Northern Railway, mines had relatively easy access to sidings on the railroad to ship ore. This greatly reduced transportation costs. Another townsite, also named Dolly Varden, was platted at the railroad siding, but it never developed. On May 20 a post office named Mizpah opened. James Herrick was the first postmaster.

The Victoria mine was once again the top producer in 1908 and 1909. During the next two years, other mines, including the Jumbo, the Copper Loop, the Tenderfoot, the Last Chance, the Keystone, the Virginia, the Edna, the Robinson, and the Dolly began production. Three mining companies, the Mizpah Consolidated Gold and Copper Mining Company, the Victoria United Copper Company, and the Dolly Varden Copper Company controlled most of the mines. However, with another dramatic drop in copper prices the mines closed one by one in 1911 and 1912, after producing $20,000 worth of copper from 1908 to 1912. The population shrank and the post office closed on November 15, 1912. By 1913 only ten people were left in Mizpah and Dolly Varden.

A small revival took place in the mid-1920s, when three mines (the Dolly Varden Queen, the Austin, and the Victoria) produced $2,850 worth of copper in 1925, but by 1927 the district had been abandoned. Another revival occurred from 1941 to 1947 when the International Smelting and Refining Company, later taken over by Anaconda Copper, reopened the Victoria mine. In 1941 the mine produced 187,000 pounds of copper valued at $22,000, but each year afterwards production slipped dramatically. In 1943 African American miners released from war duties began working at the Victoria mine, marking the first time African Americans had worked in underground mines in Nevada. The Victoria produced a total of $74,000 worth of copper from 1941 to 1947. No other ore mining took place until 1975, although in 1953 the Gulf Oil Company hired the Miracle and Wooster Drilling Company of Casper, Wyoming, to drill an oil well, the Dolly Varden no. 1. The company drilled a 10,000-foot hole at a cost of $1.5 million, but found no oil.

In 1973 Anaconda returned to work the Victoria mine. A large open pit started in 1975, but the operation was converted to underground production in 1976. A 1,000-ton concentrator began operations in 1975 and produced 20 million pounds of copper and 170,000 ounces of silver before depressed copper prices forced the mine's closure. Day Mines bought the Victoria mine from Anaconda in 1979, rehabilitated the underground mine, and began production in 1980. Between 1980 and 1981 the company produced more than four million pounds of copper. The old concentrator was unusable, and ore was shipped to McGill (in White Pine County) until depressed copper prices forced the mine to close in late 1981. No mining has taken place since.

During the 1970s and 1980s many buildings went up at the Victoria mine. As part of a reclamation agreement, all of them were removed in the late 1980s. The company sold the concentrator and moved it to Cripple Creek, Colorado, in 1988. New operations wiped out the old townsite at Watson Spring, and only a couple of dilapidated buildings mark the Mizpah townsite. At Dolly Varden, a number of foundations and other rubble remain. There are also two railroad cars at Dolly Varden, but a local rancher brought them in, and they are being used as cow sheds.

Large home still stands at Ferber. (Photo by Shawn Hall)

Ferber

DIRECTIONS: From Ferguson Springs, head south on U.S. 93A for 2½ miles. Exit left and follow the road for 2 miles. Take a right and follow the road for 11 miles. Turn left and follow that road for 2 miles to Ferber.

In 1880 the Ferber brothers discovered the Big Chief mine, the first mine in the Ferber district, but little development occurred, and their claims expired. Newton Dunyon and John Borremont took over the Big Chief. They also began working the Red Cloud, the Jigger, the Red Bird, and the Thompson mines but found little good ore. Dunyon returned to Ferber in 1883 with partners W. A. and E. B. Springer and worked the Cannonville, the Clara, and the Emma mines, while N. A. Springer began work on the Lyon mine.

In 1885 a controversy arose when G. D. Shell, C. W. Watson, Fred Snively, and George Chandler filed claims for the Mammoth and the Elko mines. These claims were already the existing Red Cloud and Big Chief mines, which Dunyon owned. His claims to the property were proven correct, but it really did not matter because the mines turned out to be worthless. A last effort was made in 1887 and 1888 when the Red Warrior, the Eva May, the Tillie, the Great Eastern, the Great Western, the R. E. Lee, the Savage, and the Emma Lou temporarily became active. Chisolm, Dunyon, and Hall also owned the Jennie Carrie, the Big Monster, and the Isaac Newton mines.

While the claims were relocated in 1890, no organized activity took place in Ferber until 1908, when G. W. and S. A. Knowlton began the Sidong mine in January, and Leffler Palmer started the Regent mine later that spring. These mines, which produced silver and lead, were the only two worked in the district until 1912.

A new strike in the Regent mine in 1912 led to renewed production of copper and iron ore, and the next five years saw a flurry of activity. A number of new mines began producing: the Salt Lake, the Martha Washington, the Knowlton, the Red Cloud, and the Ajax. The Martha Washington became the best mine of the district, producing most of the silver and lead. The Salt Lake group, owned by the Ferber Copper Company, produced the bulk of the copper, and the Deep Creek branch of the Western Pacific Railroad took ore to Erickson. In 1917, the district's best year, it produced 150,000 pounds of lead, 9,000 ounces of silver, and 60,000 pounds of copper—half of total recorded production. However, by 1918 everything suddenly shut down, and production halted. The only exception was the period between 1925 and 1940, when the district produced a small amount. Some minor shipments were made in the three years following 1940, but no mining took place after that, except in 1950 and 1957. There has been no other mining activity in Ferber since, and though claims are maintained, additional exploration has not occurred. Ferber is an interesting place to visit and to explore, because while a town never formed there, there are many mines, gallows frames, and cabins scattered throughout the district.

Ferguson Springs

(Allegheny) (Mountain)

DIRECTIONS: *Located 23 miles south of Wendover on U.S. 93A.*

George Washington Mardis, among other prospectors, first discovered copper and silver at Ferguson Springs during the summer of 1880. Mardis dubbed the area the Allegheny District, in honor of the Allegheny Mountains in his home state of Pennsylvania. After Mardis was murdered near Gold Creek in September 1880, the district remained empty except for some prospectors until after the turn of the century. In 1894 D.D. Jolly erected a 200-gallon water tank in the area and began a short-lived water company, but it proved difficult to sell water after everyone had left. Some claims were filed in 1904, and the Ferguson Springs Mining District, named for local rancher George Ferguson, formed. The same year, a school opened with five students and remained in operation until 1933.

The oldest patented claim in the district was the Red Boy, or the Danger-line. It was filed on January 1, 1910. But little, if any, production occurred until 1917. J.M. Hill visited the district in 1914 and voiced his doubts that anything profitable would be found there. Some copper, silver, and 8,000

pounds of lead were mined in 1917, but nothing else was produced until 1937. In June the Dead Cedar Mining Company formed and worked twenty claims on the west side of Ferguson Mountain. The Dead Cedar mine, formerly the Red Boy claim, produced virtually all the ore ever mined in the district. During the next three years 35,000 pounds of copper were mined, but after 1939 ore was produced only in 1949 and 1953. There has not been any mining activity in the district since. All told, the Ferguson Springs Mining District produced 45,000 pounds of copper, 8,000 pounds of lead, and 1,000 ounces of silver. An ore chute, a mine shaft, and a couple of buildings mark the Dead Cedar mine.

Kingsley

(Kinsley) (Antelope)

DIRECTIONS: *From Dolly Varden, backtrack to U.S. 93A. Head south for 2 miles. Exit left and follow the road for 9 miles to Kingsley.*

In December 1862 Felix O'Neil discovered the first lode of ore in Elko County and established the Antelope Mining District. The following year a group of renegade Mormons drove him from his mine. In 1865 George Kingsley, a former soldier at Fort Ruby, rediscovered O'Neil's claims. He began working the Emmet nos. 1, 2, 3, 4, and 5; reorganized the mining district; and named it Kingsley. By 1867 the area had aroused enough interest that a dozen men were working thirty claims, and the average return on a ton of silver was sixty-four to ninety-three dollars. A small smelter started in 1867 but quickly proved unsuccessful. While some minor production did take place, the remoteness of the location and costs of shipping ore to proper treatment facilities proved prohibitive. Newton Dunyon owned many of the mines, including the Mayflower, the Silver King, the Rocky Bar, and the Flagstaff, which were being worked in Kingsley. Other mines active in Kingsley in the early 1870s included the Chinn and Chase and the North Star. But by 1874 Kingsley had been abandoned.

In November 1878 John Murphy, Robert Cullen, and Edward Reilly discovered copper ore in the district. In May 1879 the trio, which had taken part in the earlier mining activity in the 1860s, organized the Steptoe Mining Company. In July 1881 Robert Cullen, Antoine Bixel, and Fred Kinglebock formed the Kinsley Copper Mining Company and began working a new group of claims that included the Silver, the Silex, and the Atlanta mines. Cullen also had a few mines at nearby Ferguson Springs. The name Kinsley was a corruption of George Kingsley's name, and the misnomer continues to appear on maps today. John Murphy also branched out and worked three silver mines: the Shoshone Queen, the Betsy Turner, and the Centennial. However, all these operations had failed by 1882. The last to go was the Kinsley Company. Mounting debts forced it into bankruptcy, and the company sold its property to A.C. Hormer, president of a Philadelphia-based company, for $10,000 in April 1883. There is no evidence that the company that bought the property ever conducted any mining at Kinsley.

Mill ruins at Kinsley.
(Photo by Shawn Hall)

The Morningstar mine, which John Phalan, Ed Barton, and J.P. Murphy first discovered in 1881, reopened in 1886. M.H. Lipman, D.W. Scribner, William McCrae, and T.J. Barker worked it. The company hired ten men to work the mine, which produced more than $6,600 before closing in 1887. Other active mines in Kingsley during this period included the Baltic, the Last Chance, the Elk Horn, the Ontario, the North Star, the Legal Tender, the Stone Wall, the Evaline, the Nevada, the Dolly, the Nellie, and the Eureka.

It was not until 1909 that mining activity returned to Kingsley when the Kinsley Development Company reopened the Morningstar mine and built a small concentration mill. But the company folded in 1911 after a bout of limited production. In 1913 the Kinsley Consolidated Mines Company gained control of all claims in the district and developed the Dunyon, the Phalen (formerly the Kinsley), and the Morningstar mines. The Phalen proved the most valuable of these, producing 14,000 pounds of copper, 56,000 pounds of lead, and 5,000 ounces of silver from 1914 to 1919. No additional production took place until 1930 when miners reworked old dumps and created a short-lived revival that produced bullion valued at $1,360.

Tungsten discoveries in 1944 led to sporadic mining through 1956. During this period, mines in the area produced 127 units. In December 1955 the Vesta Lee Uranium and Thorium Corporation built a rather extensive camp at the Phalen mine. The camp included six dwellings, a boardinghouse, a compressor house, a machine shop, and a treatment plant. However, after six months of conducting exploration but finding nothing, the company left, and since 1956 the Kinsley District has not produced anything. In the 1980s extensive exploration for disseminated gold took place in Kingsley. By 1988 Cominco American Resources had outlined a deposit of 2.6 million tons of ore yielding .046 ounces of gold per ton. In July 1994 Alta Gold Company broke ground for a heap leach mine at Kinsley and made the first shipment of gold in January 1995. Four open pit mines have been established. It has been predicted that they will produce 44,000 ounces per year, and total production is estimated at 157,000 ounces.

While the Kinsley District's total production is relatively low at 25,000 pounds of copper, 65,000 pounds of lead, and 5,800 ounces of silver, there is a lot to see there. At the Phalen mine, some cabins remain, and a couple of other mines also have buildings that are still standing. In the canyon to the south are the remains of the concentration mill. The Kinsley District is quite remote and far away from a gas station, so make sure you have plenty of gas before venturing there.

Lafayette Mining District

DIRECTIONS: *Located 4 miles north of Tobar.*

The Lafayette Mining District organized as a result of the discovery of silver-lead ore in 1925. The 700 pounds of ore shipped that year was all the district has ever gone on record as producing. While there are active claims in the district, no exploration has taken place. Only a small ore dump and a collapsed shaft mark the site.

Proctor

DIRECTIONS: *Located 3 miles east of Silver Zone Pass, 1 mile south of I-80.*

People began to notice Proctor in 1872 when Frank Proctor's discoveries led to the formation of the Proctor Mining District. Following a brief rush to the area, however, the veins ran out and the mines were abandoned. When the Western Pacific Railroad was completed through the area in 1909, Proctor became a stop and signal station for the railroad. This revived interest in mining in the district. The Silver Standard and Star mines, located four miles from the Proctor siding, were discovered, and because the railroad was so convenient to the mines, extensive exploration took place. However, the Western Pacific Railroad began construction on the Arnold Loop, allowing Proctor to become a small town that served as a base for the construction workers, but dooming the new town at the same time.

In March 1910 John Cardon announced his intention to build a forty-room hotel at Proctor, but when he realized that the town's life expectancy was very short, he quietly canceled his plans. The railroad completed the Arnold Loop in 1914 and everyone left Proctor. At that time, interest in mining the area disappeared. The Proctor District's only period of recorded production took place between 1917 and 1921, when the Silver Hoard and Nick Del Duke (Keystone) mines produced 38 pounds of copper and 304 ounces of silver. Only mine dumps mark the district, and nothing remains at the Proctor station.

Shafter

(Bews)

DIRECTIONS: From Wendover, head west on I-80 for 22 miles to the Shafter exit. Head southwest for 7 miles to Shafter.

Bews was established as a siding for the Nevada Northern Railway in 1906. It was named for Richard Bews of England who established a ranch there in 1897 and ran a stage and freight station. In April 1906 the railroad organized a construction camp that contained 150 Greeks and Italians. Bews was renamed Shafter when the Western Pacific Railroad reached there in September 1907. It was named after General W. R. Shafter, who was a commander in the U.S. Army in Cuba during the Spanish-American War. As the Western Pacific neared Shafter, numerous lots were sold in the town. The first store, the Morgan-Spencer Mercantile Company, opened in early 1908. Orsen Spencer became the first postmaster when the office opened in his store on August 28, 1908. The Western Pacific began regular operations on November 9, 1908.

A community of about forty lived in Shafter for years, and both railroads maintained section crews and small depots there. A school opened at Shafter in 1909 and did not close until 1933, when the county ran out of money. Schoolteacher M. J. Williams formed the Shafter Literary Society in October 1932, and the *Nevada State Herald* reported on society happenings. The society folded when Williams left after the school closed. In the early 1930s Clarence Neashum opened a general store and ran it until his untimely death in January 1941. His wife then ran the store until 1947.

In February 1950 a death occurred at the Shafter Store, which Joe Thomas ran. Dee Gower threatened Thomas, and Thomas killed him with a shotgun. In court, he testified that, "I had to shoot him. It was a case of either shoot him or let him get a hold of the gun and kill me."[1] Thomas was absolved of the murder. In the 1950s Shafter's importance greatly diminished, and in September 1953 Joe LaFrance was named postmaster, mostly because he was the only eligible resident. The office closed on April 19, 1957. All businesses closed in 1957, and Shafter was basically abandoned after that. The Railway Express Agency officially discontinued its Shafter office in March 1959. Until a few years ago, a couple of buildings still stood at Shafter, but they have since been dismantled. Numerous concrete foundations are now all that mark the site, and a small cemetery is located nearby.

Shafter School. (Jean Jepson collection, Northeastern Nevada Museum)

Not much more than foundations are left at Shafter. (Photo by Shawn Hall)

Silver Zone

DIRECTIONS: Located 6½ miles northeast of Shafter.

The pass at Silver Zone was part of the Hastings Cutoff of the California Trail. The Donner Party navigated the pass before they met their fate in the Sierras, and signs of the trail are still visible in the area. The name Silver Zone came about after Major Robert Goldman discovered silver in the area in May 1872. During the summer, a small boom camp formed, and Moffitt and Gassett began running a stage and freight line from Toano. Active mines in the district were the Silver Zone, the Wilson, the Currier and Goldman, the Poor Men's, the Star of the West, the Governor Bradley, and the Delmonico. A post office opened at Silver Zone on August 27. Charles Toyer was postmaster. But the boom had gone bust by the following summer. The camp was empty by fall, and its abandonment forced the post office to close on September 10. Some mines, including the Golden King, the Ledger, the Golden Prize, and the Little Treasure, produced in the mid-1880s.

It was not until 1907 that Silver Zone revived when construction of the Western Pacific Railroad led to the organization of a work camp. T.J. Connelly built a saloon a short distance away from the camp, but it violated the Western Pacific's three-mile limit for alcohol consumption. Officials arrested many workers and had Connelly's license revoked. When the railroad completed construction in the area, the camp disbanded, and Silver Zone became a siding and housed a section crew.

At 5,875 feet, Silver Zone pass is the highest point on the Western Pacific between San Francisco and Salt Lake City, and it has had its share of railroad accidents. The most serious of these took place in March 1936 when Engine no. 9's boiler suddenly exploded, killing brakemen Bud Howell and C.E. Dickerdoff. The two were in a caboose that the engine was pushing, and the explosion threw them through the roof. The conductor, Fred Black, was severely scalded and died five days later. Three men in the nearby section house were also badly scalded, but they survived.

In 1942 tungsten was discovered at the Silver Zone and Great Western mines, which Lester Hice, Robert Hice, and O.T. McVey sold to the Rare Metals Company of Lovelock. The company hired Robert Hice as a foreman and produced about 200 units of tungsten before leaving the district. While trains still rumble through Silver Zone pass, there is little left there except for the concrete ruins at the railroad siding. At the old mines, a few signs of the short-lived boom camp remain.

Spruce Mountain

(Sprucemont) (Spruce) (Hickneytown) (Black Forest)
(Latham) (Jasper) (Steptoe) (Johnson) (Killie)
(Monarch)

DIRECTIONS: *From Wells, head south on U.S. 93 for 36 miles. Exit left and follow the road for 5½ miles to Sprucemont. Other Spruce Mountain camps are scattered over the next six miles. They are Sprucemont, Monarch, Killie, Black Forest, and Jasper.*

Spruce Mountain, located in southern Elko County, has been the scene of mining activity since 1869. During those years, small communities named Sprucemont, Spruce, Hickneytown, Monarch, Black Forest, Latham, Jasper, Steptoe, Johnson, and Killie came to life and then declined. These camps stretched for six miles, from the western to the eastern slopes of Spruce Mountain. The area's history is fascinating, and it is one of the most intriguing sites in Elko County.

Spruce Mountain's illustrious history began in 1869, when W. B. Latham discovered the Latham mine, which was later renamed the Killie. The lead-silver ore was sufficiently valuable that a small rush of prospectors and investors flocked to the area. Within months three new mines, the Black Forest, the Juniper, and the Fourth of July, began production. By 1870 close to 100 men were working mines and claims around Spruce Mountain. In August 1871 the first large mining sale in the area occurred when Latham,

Isaac and Abner Wiseman, Thomas Kane, and Nick Shone sold the Latham mine to Crawford and Company of Philadelphia for $30,000.

Three separate mining districts were organized. They were the Latham, the Johnson, and the Steptoe districts. On September 26, 1871, all three were consolidated, and the Spruce Mountain Mining District came into being. N.H. Cotton was its president. New mining companies continued to enter the district. The two most prominent among these were the Ingot Mining Company and the Starr King Mining Company. In October the Ingot Company, which owned the Monarch and the Eureka mines, paid $80,000 for the Latham and the Fourth of July mines. The Starr King Company controlled the Badger, the Carrie, and the Grecian Bend mines.

Two separate camps formed at Spruce Mountain. Sprucemont was on the western slope, and a company town known as the Starr King property was on the eastern slope. Newspaper accounts greatly exaggerate Spruce Mountain's growth, claiming that 4,000 people lived there, making it the largest community in Elko County. In fact, however, the population was 500. There were wild claims of ore worth $4,300 per ton, but while the ore was indeed valuable, its value was nowhere near that amount. The Ingot Company produced ore worth $17,000 in 1871. The following year, under J. J. Crawford's supervision, construction began on a twenty-five ton smelter at Sprucemont. But the $80,000 smelter proved unsuccessful and closed by the end of 1872, bankrupting the company, which folded in early 1873.

In 1873 the Starr King Company began building a smelter of its own to treat ore from the Grecian Bend mine. The furnace, which cost $24,000 to

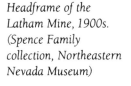

Headframe of the Latham Mine, 1900s. (Spence Family collection, Northeastern Nevada Museum)

construct, started up on May 5, just three weeks after the foundation had been laid. Mrs. J. W. Witlatch, wife of a local mine owner, shoveled the first load of ore.

By 1872 the town of Sprucemont had a population of close to 200, and Woodruff and Ennor opened the Humboldt Wells and Sprucemont toll road. The town was anything but tranquil. Rowdy miners patronized local saloons, and nights were rarely quiet. On August 29 the stage from Wells was robbed just west of town. The two robbers demanded the Wells Fargo box and were disappointed to find it contained only letters. After taking fifty dollars from the passengers, the thieves fled and were never caught. In December W. B. Langford stabbed Morris Margum at Snyder's Saloon during a drunken brawl. Margum recovered, but Langford was sentenced to the state prison in Carson City. On the other side of the mountain, the company town at the Starr King continued to grow. Its population stood at around 150. Businesses under the ownership of the Starr King Company provided goods and services.

While a considerable amount of activity was taking place around Spruce Mountain, the level of actual production was relatively low. This was discouraging for many residents, and by 1873 the district's population had dropped from a high of 500 to less than 100. The exodus hit Sprucemont especially hard, because after the Ingot Mining Company folded, the remaining active mines were all on the eastern slope. But Sprucemont clung to life, and the Schell Creek stage line had a daily stage that served the community. All supplies for the Starr King camp were shipped through Sprucemont. During the boom of 1871 and 1872 a hotel called The Globe, four saloons, and

Smelter stack and boiler at Jasper townsite. (Photo by Shawn Hall)

Post office and store at Sprucemont. (Photo by Shawn Hall)

various other businesses operated. After 1872 they closed one by one. A post office opened on April 29, 1872. Timothy Gallagher was postmaster. While Sprucemont had its ups and downs, the post office operated until August 31, 1896.

On January 22, 1874, the Ingot property was sold at public auction, but the new owners and the old ones were actually one and the same. They hoped that by purchasing their own property at a depressed auction price they would reduce the amount of money they owed on the property and would be able to resume operations. They were unable to start up again, however, and in June Ingot deeded its holdings to the Globe Smelting Company of New York City for $500,000. Many other mining deals were made in 1874. In February Porter and Cowen took over the Black Forest mine, and Bennet and Company took over the Monarch. The Starr King added to its holdings in August when it purchased E. P. Lee's Nevada Gem and Homestake mines.

A miners' strike in 1875 led to Spruce Mountain's virtual abandonment. By the end of the year, only fifty people were left. From 1875 to 1880 mainly assessment work and exploration took place in the area, and only the Starr King Company managed to sustain limited production. In February 1876 the Globe Smelting Company folded, and its property was sold at public auction in Elko for $16,900. A new company, the St. Louis Mining Company, was incorporated and began working the district in June, but it folded after finding little ore. Despite the slowdown, Spruce Mountain's need for law became apparent, and Elko County appointed Andrew McClain as constable. The first real spark of new life came in April 1880. Seven businesses were

One of the oldest buildings on Spruce Mountain, a boarding-house near the old Latham Mine. (Photo by Shawn Hall)

still active at the time, but the district's population had shrunk to fifty. The Milo mine became active and began producing seventy-five-dollars-per-ton ore. The Milo Mining Company formed, and in August constructed a small fifteen-ton furnace. This was the beginning of a strong revival, and other small mines in the area began to attract interest from outside the mining district.

In July 1883 Jasper Harrell, prominent rancher and partner in the powerful Sparks-Harrell cattle operation in northern Elko County, sold his ranches and share of the partnership for $950,000. He purchased many of Spruce Mountain's mines, including all the Starr King Company's holdings, and immediately began expanding them. A boom was on, and the rejuvenated Starr King company town was renamed Jasper in honor of the camp's new benefactor. A hotel, a store, and two saloons opened. Abner Wiseman and Warren Angel, prominent Clover Valley ranchers, owned the store. Spruce Mountain's revival was a boon for all the Clover Valley ranchers and farmers. They prospered thanks to their valley brethren's store in Jasper, and Jasper's residents enjoyed a constant supply of fresh fruit and vegetables.

Abner and Isaac Wiseman had been active in local mining since 1870 and were involved in the first sale of mining property in the area. Their store operated until 1889. During the Wisemans' years at Spruce Mountain, they bought seven parcels of lode mining property. In 1875 they paid Alexander Beaton $8,000 for 1,680 feet of quartz veins. While the Wisemans did some mining, they were primarily interested in speculation and made little money from the mines. On December 10, 1884, A.J. Roulstone finished a twenty-ton furnace for Harrell. According to his contract, Roulstone had to complete a successful thirty-day run before Harrell's company would accept the mill. Harrell hired one hundred men for the mill's construction and kept most of them on when it was completed. This greatly increased his ability to expand his operations.

Sprucemont, originally the largest community on Spruce Mountain, did not share in this revival until the Ada H. mine became active in 1886. In 1884 the town had less than twenty residents. Despite Jasper's larger size, the only post office in the district was in Sprucemont. In 1886 the Sprucemount post office was renamed Sprucemont, and people knew it by this name until its closure in 1896.

The Harrell-controlled Starr King Mining Company kept Jasper growing. By June 1885 three new mines, the Scorpion, the Friday, and the Belt, were producing ore. The furnace operated twenty-four hours a day under the management of Andrew Harrell, Jasper's son. G. W. Small served as the mining company's superintendent. The town of Jasper now included two saloons, two restaurants, a hotel, two general merchandise stores (Wiseman and Angel, Madelina and Cannon), a butcher shop, a drugstore, a livery stable, and a blacksmith shop. Ten families lived in the town. In October Jasper Harrell fell dangerously ill and had to move to California. Colonel E. P. Hardesty, prominent Elko County resident, took over company operations. By March 1886 Jasper's population had reached its peak at 175.

Other mines active near Jasper included the Juniper and the Black Forest. In April the first school on Spruce Mountain opened at Jasper. Between 1885 and 1888 mines in the area produced almost $200,000 worth of primarily copper, mostly from Starr King properties. Other mines also produced fair amounts. The Jumbo mine generated revenue in the amount of $40,000 while the Ada H., Sprucemont's mainstay, brought in $11,000. Two hundred miners lived on Spruce Mountain by June 1887, and five saloons in town served the thirsty men. The mines were still desperately short of men, though, and they offered higher-than-normal wages in order to induce more people to come work for them. Miners received $6 a day, smelter workers $5, feeders $4, and helpers $3.50. The Juniper Consolidated Mining Company completed a new furnace, but it was unsuccessful and closed down within a year. By the end of 1888 the mines began to fail as ore veins thinned and faded. The smelter in Jasper shut down in 1889, and by the end of the summer all mining had stopped. The ore on Spruce Mountain was very porous, and smelting techniques were not efficient enough to profitably extract silver from the poorer-quality ore. When the Starr King Company folded in 1890, Spruce Mountain entered an extended period of very low activity.

Mostly insignificant production occurred on Spruce Mountain until 1899, when the Monarch Mining Company was organized. This was the beginning of Spruce Mountain's longest period of sustained activity. Another mine, the Hartley, began production, and a new gas hoist was installed in April 1900. By the end of 1900 Spruce Mountain's population had risen to about seventy-five. In 1901 Charles Spence, who organized the Black Forest Mining and Smelting Company, reopened the Black Forest mine. By October he had started a smelter at Jasper. In March 1901 A. J. Reed, nicknamed "Copper King" Reed, reopened and cleaned out the Jasper (formerly Starr King) mine. Two mines reopened on the Sprucemont side, and a couple of businesses opened. One of these was the Sprucemont Hotel, which Mrs. Healy ran.

In April there was a new strike on Banner Hill, and the Four Metals Mining Company reopened the Latham mine. In July the Latham Mining Company was organized and took over the Latham mine, where it built a concentrator. Most company employees resided in Jasper. The company contracted with J. J. Wiseman, Frank Avery, John Williams, and Frank Smith to haul coke and bullion to and from Wells. In July 1904 the Black Forest smelter closed after producing 700 bars of silver. The ore lacked sufficient sulfur content for smelting, but the smelter was refurbished with better equipment, enlarged to thirty tons, and restarted in October. It was enlarged again, this time to forty-five tons, in 1906. Sprucemont did not benefit much from Jasper's revival, and most Sprucemont residents moved to Jasper. In March 1905 Mary Hamilton sold Sprucemont's last hotel, shop, and stable for $400. Lumber from the buildings was used to construct boardinghouses at Black Forest.

In August the Ohio Lead Mining and Smelting Company took over the Latham Mining Company and began working the Latham and Juniper mines. Robert Hartley sold the Hartley, the Index, the Bengal, and the Spring mines to the Ohio Company for $14,000, but the Black Forest Company was still the most prominent in the region, producing virtually all the lead and most of the silver mined in Elko County. A new wagon road ran to Bews, or Shafter, on the Nevada Northern and Western Pacific Railroads. In November George Talbot, Nevada Supreme Court justice, sold the Killie and the Junior mines to Montana capitalists. In 1907 the Big Four Nevada Mining Company was organized. Frank Arford was president, and William Shelton was manager. The company's mines were located on Banner Hill and included the Neversweat and Keystone mines. The ore was smelted at the Black Forest smelter.

Another new company, the Spruce Mountain Copper Company, which William Dunn, Carlton Hand, and A. A. Jones owned, soon joined the Big Four. They consolidated their holdings, including the Contact mine. While silver was still Spruce Mountain's main product, people began to become aware of the value of copper, lead and zinc, and made new copper discoveries. The Spruce Mountain Company had a major strike in August and became the district's biggest producer. It bought all the Ohio Lead company's holdings, controlled most of the producing mines, and absorbed the Black Forest Company and smelter. The company named Charles Spence manager of all its Spruce Mountain operations. In August something strange happened to miner William Vineyard while he was working on the 200-foot level of the

Spruce Mountain Company's new discovery. A fierce thunderstorm began above ground, and a bolt of lightning came down the shaft and hit the rails on which he was standing. He was badly shocked and bruised, but lived through the experience.

The financial panic of 1907 had a devastating impact on companies active in the district, and while some continued exploration, only the Spruce Mountain Copper Company shipped any ore in 1908. In October William Shelton, manager of the Big Four Company, died suddenly in Chicago. The company left the district shortly afterward. The smelter at Jasper closed in 1908, and little production occurred during the next few years. After eighteen years on Spruce Mountain, Spence finally found a rich deposit in April 1911 in the Latham mine. In May another strike took place on the Great Eastern Mining Company's property. These discoveries helped revive interest in the area.

In February 1912 William Johnson, F. F. Dalton, and James Davis formed the Bullshead Mining Company. The company acquired the old Harrell holdings that had been abandoned since 1890 and began shipping ore to Garfield, Utah for treatment. Another new company, the Banner Hill Copper Mining Company, began operations in March 1913, but all the companies' production levels were low, and by September they were employing only ten men. In 1914 the Spruce Mountain Copper Company folded. The district was basically inactive and unproductive until 1916, when Charles Spence regained control of his old properties. He built a fifty-ton furnace on the edge of the valley to the east of Jasper to treat ore from his Black Forest and Bullshead mines. In 1916 the Keystone, the Ada H., and the Black Forest mines were active but only produced $13,000 worth of primarily copper. Spence incorporated his Black Forest Mining Company in February with an optimistic $1 million in stock, but production was slow until 1918.

Mineral prices, boosted by World War I demand, led to increased activity. Bart Woodward and Herbert Badt, both of Wells, took over the Spruce-Monarch mine. The Spruce Mountain Monarch Mining Company, which also operated the Ada H. mine, formed. Herbert Badt was president and David Neff vice-president. The Monarch mine, located on the western slope of Spruce Mountain, was the deepest in the district and Woodward, formerly from the Tintic District in Utah, was named its manager.

Judge C. E. Mack and William Johnson bought the Bullshead property, and the New Bullshead Mining Company purchased the old Salmon River Mining Company's smelter in Contact. This smelter's specialized treatment process could easily process the difficult Spruce Mountain ore. Captain Herman Davis was in charge of the Bullshead mines and smelting construction, and soon there were twenty buildings on the Bullshead property, located at the old townsite of Jasper. In addition, the company built a number of other structures at the mine site, located at the head of the canyon just south of Jasper.

The only three producers in the area in 1918 were the Spruce-Monarch, the Bullshead, and the Black Forest mines, which produced primarily silver and copper valued at $23,800.

The period of the district's greatest production, which lasted more than twenty years, started in 1919. In November the Bullshead Company completed its fifty-ton smelter, which produced twenty-four 100-pound bars of silver in its first week of operation. By the end of 1919, the Bullshead Mining Company, the Spruce Mountain Monarch Mining Company, the Spruce Mountain Consolidated Mining Company, the Black Forest mine, and the Breedin and Riter properties were all producing. Ore was shipped through railheads at Currie and Tobar. The Spruce-Monarch mine became the primary producer, and in 1920 it yielded 7,000 tons of silver-lead ore. In January the company had a major strike in the mine, which Fordham and Stratton originally discovered in 1901. Six large motor trucks hauled ore to Tobar for shipment to Salt Lake City. A new company, the Ada H. Mining Company, formed in March. Samuel Neff was president and Herman Badt vice-president. Bart Woodward, who was still mining around Spruce Mountain, was named manager. The ownership was basically the same as that of the Spruce Mountain Monarch Mining Company, but the two were kept separate for business purposes. In May Charles Spence, who had worked the Spruce Mountain mines since 1887, passed away at age sixty, but despite his death the Black Forest mine continued to expand and to increase production.

The companies active in the district started to become aware of the multitude of minerals present in Spruce Mountain ore. Besides silver, lead, and copper, they were also finding manganese and zinc. In 1922 the Bullshead Company began shipping manganese ore from its J. B. Wall lease. The value of Spruce Mountain's production in 1922 and 1923 stood at $100,000, but better times were just around the corner. In 1924 total production value stood at $113,000, which included one million pounds of lead. The Spruce-Monarch mine was the best in the district. With active mines on both sides of Spruce Mountain, Sprucemont and Jasper revived. In addition, two new camps formed at the Black Forest and Spruce-Monarch mines. Spruce Mountain's population grew to 150. The Spruce Mountain Monarch Company took over the Black Forest property and reopened the Killie mine. By August it employed eighty people and had five active mines.

One of Spruce Mountain's most prominent mine managers, Bart Woodward, who was also part owner of the Polar Star mine at Warm Creek, was seriously injured when the team he was working backed over the edge of the ore dump at the Index mine. His recovery was long and painful, but he eventually returned to his mines at Spruce Mountain. In 1924 the Paramount Consolidated Mining Company purchased the Ada H. mine. In 1925 Andrew Harrell returned to Spruce Mountain to work some of his father's old mines. Because of Spruce Mountain's increased population, a post office opened at Black Forest on January 9, 1926, with Henry Crittenden as postmaster. Another post office opened at Sprucemont on November 1, 1929. From 1926 to 1928 seven mines produced $400,000 worth of bullion. The Bullshead Mining Company gained control of the O'Neill mines in April 1926, creating a construction boom at its company town on the old site of Jasper. The first dances at Spruce Mountain since the 1880s took place at the boardinghouse, and the *Nevada State Herald* declared that "old Spruce Mountain is coming back!" The Russell Freighting Company built four homes, a truck garage, a warehouse, and a double compartment ore bin. The company ran three trucks to Jasper Station on the Western Pacific Railroad. From there the ore was sent to the American Smelting and Refining Company in Murray, Utah. A school also opened on Spruce mountain, and there were six students.

In August 1926 Nevada Governor Scrugham visited Spruce Mountain as Herman Badt's guest. There were six major mining companies in the district; they were the Spruce-Monarch Lead and Silver Mining Company, the Black Forest Mining and Smelting Company, the Spruce Consolidated Mines Company, the Mines Compact, the Spruce Standard Mining Company, and the Bullshead Mining Company. In January 1927 the Nevada Lead and Zinc Mining Company, which W. F. Snyder and Sons owned, bought the Spruce Consolidated Company. All Bullshead and O'Neil holdings were organized under new ownership, and the Bullshead Consolidated Mining and Milling Company spent $100,000 to purchase the holdings. In May 1927 a 6,800-foot tramway serving the Spruce Consolidated Company and Black Forest properties was completed. The tram ran 3,900 feet from the Killie mine to an ore bin at the Black Forest mine. The ore was then loaded onto a second section of the tramway, which ran another 2,900 feet down the steep canyon to a truck-loading station. The tramway ran buckets that held 750 pounds each, and the line could haul fifty tons in eight hours.

In 1928 the Missouri-Monarch Mining Company bought the Monarch, the Black Forest, and other holdings of the Spruce Mountain Monarch Company. The two mines had been connected in 1926 by the 7,000-foot Bronco Tunnel, and the company, which controlled thirty-four claims and two miles of the main strike fissure, was listed on the Salt Lake Stock and Mining Exchange. The company was a consolidation of the Spruce Mountain Monarch and Big Missouri companies. Large bunkhouses were built at the Black Forest and Monarch mines. Lead continued to be Spruce Mountain's mainstay, and area mines produced one million pounds of it from 1924 to 1929. In 1929 the company mainly conducted exploration and defined ore deposits. Spruce Mountain mines produced more than $70,000 worth of ore, but this came mainly from the Nevada Lead and Zinc company's Killie mine. But Nevada Lead and Zinc folded in early 1930 because of low lead prices.

In November 1929 an investors' group filed an application to build a railroad spur from Ventosa, on the Western Pacific Railroad, to Spruce Mountain.

The trains would have hauled ore from Spruce Mountain and the nearby Mud Springs Mining District. At the same time, the spur would have provided shipping points for ranchers and farmers in the Clover and Ruby valleys. During the next year, the company had a railroad bed surveyed, but it ran from Tobar, not Ventosa, and the cost was estimated at $300,000. However, trucking eliminated the need for a railroad, and in August 1932 the Missouri Monarch Company received permission to construct a new 3,000-foot truck road from its mines to Sprucemont.

A number of new companies entered the district in 1929. The Eastern Nevada Exploration Company, which David and Samuel Neff, Bart and Charles Woodward, and Herbert Badt owned, mined the Junction claim group. Charles Woodward, a director of the Spruce Standard mine, was also president of the new Bingo Consolidated Mining Company, which owned property between the Killie mine and the Missouri-Monarch holdings. More exploration and organization of new companies took place in the early 1930s. The Centennial Consolidated Mining Company located good ore in December 1931. Unfortunately, in 1932 Spruce Mountain lost two of its biggest investors and promoters when both Herbert Badt and Bart Woodward died. The Missouri Monarch Company absorbed most of their holdings. The Parker brothers built a small mill at the Humbug mine. In 1934 the Index Mining Company was incorporated. Its directors included Charles Woodward, Samuel Neff, and R.W. Edwards. M.H. Woodward, another brother, was in charge of the office at Spruce Mountain, and besides the Index mine, the company also ran the Pal Number One and Number Two. In July 1935 the Payette Mining Company and the Branson Gold Mines Company formed. Charles Woodward was president of both.

While all this activity was going on, businesses were operating at Black Forest, Sprucemont, and Monarch. At Black Forest and Monarch, virtually all businesses were under the company umbrella. Two schools, at Black Forest and Sprucemont, were in operation. At Black Forest, Ruth Lyon of Metropolis taught seven students, while in April 1932 teacher Ula Reed Van Diver organized the first National Bank of Health program at Sprucemont. Students received credits for cleanliness, neatness, and proper eating and sleeping habits. The Nevada state nurse, Ebba Bishop, was so impressed with the program that she instituted it elsewhere in Nevada. The American Red Cross and the Cleanliness Institute of New York City published reports on the program. Franklin Delano Roosevelt's national school health program made use of many of the National Bank of Health program's facets. Despite national accolades, the school district officially dissolved in October 1933, though schools still operated periodically for a few more years.

Most mining at this time was taking place on the east side of Spruce Mountain at the Black Forest and the Bullshead mines, and Sprucemont, on the west slope, began dying. The majority of the residents left, and the post office closed on October 17, 1935. After 1937 the Missouri Monarch Company was the only sizable producer in the district, though the Index Company, which ran the Bullshead and O'Neil properties, made some minor shipments in 1937 and 1938. Lessors worked other mines such as the Rainbow, the Kelly, and the O.D., but they produced little.

Tragedy struck in November 1940 when lessor Jake Wiskerchin was trapped and smothered to death in the Monarch mine. He was working the mine with Joe Hawkins and Blaine Whimpy. Whimpy found some good ore and asked Wiskerchin to look at it. Wiskerchin swung a pick at the ledge, and two tons of dirt and rock fell, covering him from the neck down. He suffocated before his friends could extricate him. His wife and four month-old-baby had just joined him on Spruce Mountain.

While the post office at Black Forest closed on March 19, 1943, the mining companies still employed more than thirty men. The closure of the post office marked the end of many years of postal service for Black Forest, Spruce, Spruce Mountain, and Sprucemont.

Spruce Mountain's biggest year of production ever was 1945. During that year, the area's mines produced almost $300,000 worth of ore. From 1944 to 1948 the figure was close to $900,000, but by 1947 the ore began to decrease in value, and mines yielded ore valued at only $41,000. In July 1947 the Missouri Monarch Company, the Spruce Standard Mining Company, the Missouri Monarch Extension Mining Company, and the East Nevada Exploration Company consolidated to form the Nevada Monarch Consolidated Mines Company, which controlled all mines except the Killie. In September R.H. McClintock and Company spent $50,000 to conduct diamond drilling in an attempt to locate new deposits, but little production occurred during the next few years. The drilling led to a few promising leads. The final blow came when the company president, W.M. Archibald, died in Toronto, Canada. The company folded in 1952 after producing less than $30,000 worth of mainly copper and silver.

Spruce Mountain mines produced every year from 1899 to 1952. When the Nevada Monarch Company folded, serious mining on Spruce Mountain came to an end. Some leasing activity occurred through 1961, but it met with only limited success. In March 1956 the Index Daley Mining Company of Idaho (which also owned forty-eight claims and leased the Eugene Parker property) reopened the Bullshead property and employed five men. Charles Woodward, who had worked Spruce Mountain mines for forty years, wanted one more chance at success and served as president and manager of the company. During the next two years, the company undertook 1,800 feet of tunneling at Bullshead, but it found no rich deposits, and prices did not warrant shipping low-value ore. By the fall of 1958 Woodward had given up, and the company left Spruce Mountain.

Since 1961 no production has taken place on Spruce Mountain. Some exploration occurred through the 1980s, as is evidenced by two buildings filled with ore specimen bags near Sprucemont. Companies such as Newmont Exploration, Freeport, and AMAX have found deposits, but none have been worthy of extensive mining. The total value of production from mining on Spruce Mountain is $2.9 million and includes one million ounces of silver, 22 million pounds of lead, 3.2 million pounds of zinc and 780,000 pounds of copper.

Spruce Mountain is the best ghost town in Elko County. At Sprucemont, the old log post office is the only building standing, but many dugouts, foundations, and rubble clearly show the town's layout. The meticulously built stone foundations of Spruce Mountain's first smelter are located just to the north. Just east of Sprucemont is a small complex of buildings that the owners of the Ada H. mine built. About two miles farther east are the remains of the Monarch mine, where there are at least a dozen buildings in various stages of decay. Buildings from the 1890s struggle to stand next to structures from the 1940s, which is quite unusual in Nevada's ghost towns. Cement foundations and ruins of the smelter are located at the mouth of the still-accessible Monarch mine. Another mile east, on top of Killie Pass, are the remains of the Killie (Latham) mine, the original discovery on Spruce Mountain. During the 1920s the camp had half a dozen buildings. While only scraps of wood and brick mark its site, a number of log structures built before the turn of the century are nearby. Between Killie and Black Forest, one mile east heading down Black Forest Canyon, are the only remains of the tramway. The line was sold for scrap and was dismantled, except for two transoms that still stand.

Black Forest has much to offer visitors. A number of buildings, including the post office/boardinghouse, are still there, but between June and September 1993 most of that building collapsed. Two mining buildings remain. One housed a small smelter, while the other contains original equipment used to pump water out of and fresh air into the mine. About half a mile past the mine are another group of frame buildings that comprised the main Black Forest camp. The barely erect frame building housed the mining offices, and a boardinghouse is next door. Across the road are the cement foundations of the school and the ruins of family dwellings.

Another mile east is Jasper, where buildings are scattered along the quarter-mile townsite. Most are miners' quarters built during the Bullshead operations in the teens and twenties. A couple of pre-turn-of-the-century cabins still stand. Foundations of two smelters are visible, and the metal stack of the Bullshead smelter remains. At the Bullshead mine, located high in the adjacent canyon to the south, are extensive mining ruins. Some stone ruins from the 1870s exist amid rubble from buildings constructed later. The last signs of operations at Spruce Mountain are located about three miles east of

Jasper, at the edge of a valley where only slag heaps and one wall mark the site of the smelter Charles Spence built in 1916.

Besides the buildings and townsites, there are a multitude of mines, head frames, ore chutes, and dumps at Spruce Mountain, where the old and the new exist side by side. Spruce Mountain provides Elko County's best opportunity to view the passage of a hundred years of history.

Tobar
(Clover City)

DIRECTIONS: *From Wells, head south on U.S. 93 for 14 miles. Exit left and follow the road for 4 miles to Tobar.*

The Western Pacific Railroad initially established Tobar as a construction camp in 1908. The railroad built a substantial depot there, and a small town began to form. During the construction, the owner of the Rag Saloon, which was housed in a canvas tent, put up a sign that said simply, TO BAR. The two words became one, and the new town had its name.

While Tobar was basically a railroad town, it rose to prominence when it became the center of a planned dry farming community. The first dry farmers in Tobar were Sidney Curtiss and Gover, Fred, and Ernest Wood. By 1910 there were sixteen homesteads around Tobar. By 1911 Tobar had attracted enough attention that there were about seventy-five people and twenty dwellings in the town and surrounding area. A school opened at the Munson Ranch near Tobar, and it moved to Tobar in 1913. A post office opened in town on December 20, 1911. Mack Backstead was postmaster. In 1912 the railroad decided to sell land at Tobar for four to eleven dollars an acre. The low prices enticed even more people to come to the town. In June local attorney Frank Spear slated the *Tobar Eye-Opener* for publication. However, no issues have even been located, and it seems as though the paper was never published.

Besides being a dry farming area, Tobar also became the main shipping point for ore from Warm Creek and Spruce Mountain mines and for Clover Valley produce. In 1913 the Tobar settlement really took off when A.B. Hoaglin, who platted a townsite and built a two-story hotel, later called the "White Elephant" by local residents, began a high-powered promotional campaign for the farming area. Advertisements in papers throughout the country attracted many people to the town, but the information they included was false. Brochures heralded Tobar as the home of the "big red apple," though there were no orchards in the area. The promotional campaign also claimed that 50,000 acres would be under cultivation at Tobar and that the town

would soon have a population of 3,000, but it failed to mention from where the water to irrigate 50,000 acres was to come. Officials charged Hoaglin with mail fraud, because he was selling public land, but he was later acquitted on a technicality. He then moved to Canada, but later returned. During his absence, his brother carried on the promotion.

In July 1915 the Hoaglin brothers opened the Chicago Store—a complete grocery store. Two other stores, the Golden Rule and the United, also opened in 1915. By the end of 1915 fourteen businesses were running at Tobar, all relying on patronage from the Tobar Flat farmers. Out on Tobar Flat a number of little hamlets had formed, each containing ten to twenty-five people. The population of the Tobar area was 500, of which 400 lived on farms. Half of Tobar's residents were from northern Utah. Most were former railroad workers, and their average age was thirty-nine. Residents cultivated three thousand acres, but droughts beginning in 1915 doomed any future the Tobar farms may otherwise have had.

By 1916 Tobar had begun to decline. A fire destroyed the Chicago store in March, and it was never rebuilt. On July 22 A. L. Covert published a short-lived newspaper, the *Tobar Times,* but it folded in August. Subscribers who had paid for a full year were extremely upset with Covert, who quietly left town without reimbursing them. Another paper, the *Tobar Sentinel,* which the elderly Sunday-school teacher edited, ran for a short period in 1916.

The dry farms' problems continued to increase. An extended drought made for tough times, and a jackrabbit invasion exacerbated the situation. Only one inch of rain fell at Tobar in 1916, and the rabbits ate what little grew. One farmer commented that the best crop he had was two boxes of arrowheads. Many families began to move away. Those who stayed had to seek employment in nearby mines to make ends meet.

The town received a boost when the railroad built a 50,000-gallon water town and section gang houses, but the exodus was on and farmers continued to leave. By the end of 1917 they had abandoned virtually all the dry farms. A new promotional campaign began in 1918, and as part of an effort to portray the area as lush, the townspeople had the post office renamed Clover City on December 11, 1918. Tobar's streets were re-bladed, and the Elko Valley Estates formed. Two lots were advertised at $395 each, but the number of jackrabbits squelched any interest in farming around Tobar, and all efforts fell flat. It was difficult to convince people that farming in an area that had no water and no rain was a good idea, and by 1920 dry farming was over in Tobar. Only six of the original homesteads remained occupied, and the people who were still there worked elsewhere, not on their farms. The town of Tobar struggled to survive, with its main source of income coming from ore and cattle shipments. A new store, the Sawyer Mercantile Company, opened in March. The completion of Highway 93 meant that cattle were now shipped by truck. That, combined with a depression of metal prices that slowed ore shipments, spelled the end of Tobar's days. On January 18, 1921, the post office was once again renamed Tobar. In September 1923 the Tobar Lumber Company, which E. E. Glaser had opened during Tobar's early days, moved to Wells and was renamed the Western Hardware and Lumber Company.

By the end of the 1920s only the station, the school, the Clover Valley Store, and the post office were open in Tobar. There were, however, many abandoned homes and businesses, including the old real estate office. In the 1930s mainly railroad workers and their families used what was left of Tobar. The county sold the school, which was in a serious state of disrepair, for ten dollars in February 1937. The Tobar post office closed on September 17, 1942. In 1946 the Western Pacific abandoned the depot, and Tobar's last business, the Clover Valley Store, closed. In October 1948 Hal Bricker of Wells bought the school and teacherage and moved them to Wells. One by one, the other buildings were sold and moved. By 1950 only the water tower, the depot, the ore loading chute, the cattle pens, and the section houses were left. In 1952 all the remaining railroad buildings were sold and moved. The Noonday Mines Company moved the Tobar Depot to Battle Creek, where it served as a boardinghouse for a mining operation. The station still existed until August 1995, when a brush fire destroyed the building, making it a moot point to move it to Elko, where it was to have served as the offices for the Elko Chamber of Commerce. Once the water tower was dismantled later in 1952, nothing besides foundations were left at Tobar. Tobar made the news one more time in 1969, when on June 19 a train carrying carloads of bombs destined for Vietnam exploded one mile to the west of the town, leaving a huge crater and injuring conductor T.M. Johnson and brakeman Freeman Stephens. Today, extensive foundations and collapsed cellars show the layout of the once-bustling town, but not much else is left. About one mile south of town are the concrete walls of one of the larger homesteads from the teens. A few other smaller homestead ruins are scattered on the flat below the Tobar townsite.

Valley Pass

DIRECTIONS: *Located 5 miles west of Cobre.*

The railroad created Valley Pass, a signal station on the Southern Pacific Railroad, after the completion of the Lucin cutoff. The railroad moved the Toano depot to Valley Pass and built section houses and a large water tower. Considering the amount of traffic on the sidetrack, relatively few accidents occurred there. One fatal incident did take place in July 1939. Brakeman Harry DeYoung was decapitated when he fell under the engine tender while attempting a flying switch—jumping off a moving train in order to move the track switch.

The arrival of diesel engines in the 1940s made the Valley Pass services obsolete, and the buildings were removed to other locations. Today, trains still run through Valley Pass. Just the foundations of the buildings and the impressive water tower, one of only a few left in Elko County, guard the site.

Depot foundations at Ventosa. (Photo by Shawn Hall)

Ventosa

DIRECTIONS: *Located 5 miles east of Tobar.*

Ventosa was a stop on the Western Pacific Railroad and the next depot east of Tobar. The depot and stop were used primarily as a shipping point for Spruce Mountain ore, and a special siding was built to load the ore. A school was opened at Ventosa in 1912, with G. E. Brown as teacher, and it remained open until the 1920s. From the teens through the 1940s about ten people, most of whom worked for the railroad, lived at Ventosa. In 1929 a plan to build a railroad spur to Spruce Mountain with Ventosa as the terminus attracted a considerable amount of interest, but trucks came into vogue before the line could be built, and the railroad never materialized.

In the early 1950s the depot and section buildings were abandoned, sold, and removed. Ventosa made the news one more time in August 1950 when a broken draw bar derailed thirty-two cars there. A fire broke out, and all the cars were destroyed at a loss of $250,000. Only concrete foundations are left at Ventosa today.

Wardell

DIRECTIONS: *From Wells, head east on I-80 for 8 miles to the Moor exit. Head south for 2 miles to Wardell.*

Wardell was a short-lived mining camp in the hills about two miles south of Moor. L. H. Wardell discovered the Wardell mine in 1879. He removed small amounts of silver ore but ignored the iron ore he had to go through to get to the silver. Tragedy and the devaluation of silver doomed the camp of Wardell. In October 1880 a fire at the mine killed two miners, Godfried Ammon and Jacob Hamilton, and the shaft house and most of the mine timbers burned. After that, the miners became superstitious and would not work in the mine. Wardell was forced to close the mine and abandon the district.

It was not until July 1905 that Richard O'Neil and George Vardy relocated the Wardell mine and began to produce iron ore. Moor became the shipping point for the mine. In 1909 the two began working a new claim, the Union, which initially returned $450 per ton in silver. Soon afterward, they abandoned the Wardell mine and concentrated their efforts on the Union mine. Vardy and another partner, P. Angelo, began developing the Don Pedro mine in 1913, but the veins in these mines were very short, and further development could not locate additional deposits. The men gave up, and the Wardell District was abandoned for good. There has not been much interest in Wardell since. Collapsed shafts, mine dumps, and some collapsed cabins mark the site.

White Horse

DIRECTIONS: *Located 5 miles south of Ferguson Springs.*

White Horse is an obscure mining district in which lead, zinc, and silver ore found just after the turn of the century attracted some prospectors. The district was named for the great Gosuite chief, White Horse, who triggered the Overland War of 1863 by attacking Eight Mile Station, located a few miles away in White Pine County.

Some additional work took place on claims during the teens, but the area saw little, if any, production. The only recorded production in the White Horse District took place during the 1940s. During this period, the Kathleen and the Blue mines became active. They produced 6,200 pounds of lead, 160 ounces of silver, 800 pounds of zinc, and 28 units of tungsten between 1942 and 1949. No other activity had taken place in the district since 1949, and only a shaft and old prospect pits are left at White Horse.

References and Notes

Northwestern Elko County

Aura

REFERENCES

Carlson, Helen S. *Nevada Place Names: A Geographical Dictionary.* Reno: University of Nevada Press, 1974.

Elko Independent, 16 February 1911.

National Archives. Postal Records of Nevada. 1860–1970.

Paher, Stanley. *Nevada Ghost Towns and Mining Camps.* Las Vegas: Nevada Publications, 1970.

Patterson, Edna, Louise Ulph, and Victor Goodwin. *Nevada's Northeast Frontier.* Reprint, Reno: University of Nevada Press, 1991.

Beaver Mining District

REFERENCES

LaPointe, Daphne D., Joseph V. Tingley, and Richard B. Jones. *Mineral Resources of Elko County, Nevada.* Nevada Bureau of Mines Bulletin no. 106. Reno: Mackay School of Mines, University of Nevada, 1991.

Smith, Roscoe M. *Mineral Resources of Elko County.* USGS Open File Report no. 76–56. Washington, D.C.: Government Printing Office, 1976.

Blue Jacket

REFERENCES

Elko Daily Free Press, 14 August 1907.

Elko Independent, 12 April 1885, 15 September 1885.

Tuscarora Times-Review, 22 December 1889, 14 December 1872.

Bootstrap

Elko Daily Free Press, 12 December 1993.

LaPointe, Daphne D., Joseph V. Tingley, and Richard B. Jones. *Mineral Resources of Elko County, Nevada.* Nevada Bureau of Mines Bulletin no. 106. Reno: Mackay School of Mines, University of Nevada, 1991.

Patterson, Edna. "Thar's Gold in Them Hills—Parts 3 and 4." *Northeastern Nevada Historical Society Quarterly* 4 (winter 1972): 2–18.

Bull Run
REFERENCES

Elko Free Press, 13 May 1901, 6 August 1902.

LaPointe, Daphne D., Joseph V. Tingley, and Richard B. Jones. *Mineral Resources of Elko County, Nevada.* Nevada Bureau of Mines Bulletin no. 106. Reno: Mackay School of Mines, University of Nevada, 1991.

Robertson, Opal Curieux. "William and Catherine Johnson Family." *Northeastern Nevada Historical Society Quarterly* 2 (1991): 51–74.

Tuscarora Times-Review, 18 October 1899.

Burner
REFERENCES

Elko County Recorder. *Miscellaneous Records.* Vol. 1.

Emmons, W. H. *A Reconnaissance of Some Mining Camps in Elko, Lander, and Eureka Counties, Nevada.* USGS Bulletin no. 408. Washington, D.C.: Government Printing Office, 1910.

LaPointe, Daphne D., Joseph V. Tingley, and Richard B. Jones. *Mineral Resources of Elko County, Nevada.* Nevada Bureau of Mines Bulletin no. 106. Reno: Mackay School of Mines, University of Nevada, 1991.

Columbia
REFERENCES

Elko County Recorder. *Miscellaneous Records.* Vols. 2, 10. January 1880, 8 May 1902.

LaPointe, Daphne D., Joseph V. Tingley, and Richard B. Jones. *Mineral Resources of Elko County, Nevada.* Nevada Bureau of Mines Bulletin no. 106. Reno: Mackay School of Mines, University of Nevada, 1991.

National Archives. Postal Records of Nevada. 1860–1970.

Tuscarora Times-Review, 23 August 1871.

Cornucopia
REFERENCES

Angel, Myron, ed. *History of Nevada.* Oakland: Thompson and West, 1881. Reprint, Berkeley: Howell North Books, 1958.

Ashbaugh, Don. *Nevada's Turbulent Yesterday: A Study in Ghost Towns.* Los Angeles: Westernlore Press, 1963.

260   *References and Notes*

Couch, Betrand F., and Jay A. Carpenter. *Nevada's Metal and Mineral Production, 1859-1940*. University of Nevada Bulletin vol. 37, no. 4. Reno: University of Nevada, 1943.

Elko County Recorder. *Miscellaneous Records*. Vol. 1. 5 March 1890.

Elko Independent, 22 November 1873, 11 September 1875, 18 July 1874, 30 October 1875, 17 August 1876, 17 April 1881, 20 June 1881.

Hannon, Genivieve. "Cornucopia: Elko County Silver Camp." *Northeastern Nevada Historical Society Quarterly* 2 (spring, Summer 1982): 63–69.

LaPointe, Daphne D., Joseph V. Tingley, and Richard B. Jones. *Mineral Resources of Elko County, Nevada*. Nevada Bureau of Mines Bulletin no. 106. Reno: Mackay School of Mines, University of Nevada, 1991.

National Archives. Postal Records of Nevada. 1860–1970.

Patterson, Edna. "Thar's Gold in Them Hills—Parts 3 and 4." *Northeastern Nevada Historical Society Quarterly* 4 (winter 1972): 2–18.

Tuscarora Times-Review, 12 September 1883.

State Mineralogist of Nevada. Annual Report. 1875.

U.S. Bureau of the Census. 1880.

Dinner Station
REFERENCES

Elko Independent, 12 May 1870, 24 October 1884.

Hendershot, Carol. "Dinner Station." *Northeastern Nevada Historical Society Quarterly* 3 (summer 1985): 63–79.

Patterson, Edna. "Thar's Gold in Them Hills—Parts 3 and 4." *Northeastern Nevada Historical Society Quarterly* 4 (winter 1972): 2–18.

Thompson, Marie Davidson. "The Davidson Brothers, Jack and Walt." N.d. Northeastern Nevada Museum manuscript collection, Elko, Nevada.

Tuscarora Times-Review, 1 October 1884, 6 November 1886, 16 September 1890.

NOTES

1. *Elko Free Press,* 28 June 1915.
2. Daphne D. La Pointe, Joseph V. Tingley, and Richard B. Jones, *Mineral Resources of Elko County, Nevada,* Nevada Bureau of Mines Bulletin no. 106 (Reno: Mackay School of Mines, University of Nevada, 1991).
3. Carol Hendershot, "Dinner Station." *Northeastern Nevada Historical Society Quarterly* 3 (summer 1985): 63–79.
4. Marie Davidson Thompson, "The Davidson Brothers, Jack and Walt," n.d., Northeastern Nevada Museum manuscript collection, Elko, Nevada.

Divide
REFERENCES

Elko Free Press, 28 June 1915, 4 September 1915, 23 September 1928.

LaPointe, Daphne D., Joseph V. Tingley, and Richard B. Jones. *Mineral Resources of Elko County, Nevada*. Nevada Bureau of Mines Bulletin no. 106. Reno: Mackay School of Mines, University of Nevada, 1991.

Edgemont

REFERENCES

Clawson, Marion. "History of the Clawson and Thompson Families." 1986. North-eastern Nevada Museum manuscript collection, Elko, Nevada.

Couch, Betrand T., and Jay A. Carpenter. *Nevada's Metal and Mineral Production, 1859–1940*. University of Nevada Bulletin vol. 37, no. 4. Reno: University of Nevada, 1943.

Elko County Recorder. *Miscellaneous Records*. Vol. 1.

Elko Free Press, 20 February 1904.

Elko Independent, 12 May 1904, 30 May 1937, 4 September 1938.

LaPointe, Daphne D., Joseph V. Tingley, and Richard B. Jones. *Mineral Resources of Elko County, Nevada*. Nevada Bureau of Mines Bulletin no. 106. Reno: Mackay School of Mines, University of Nevada, 1991.

National Archives. Postal Records of Nevada. 1860–1970.

Patterson, Edna B., Louise A. Ulph, and Victor Goodwin. *Nevada's Northeast Frontier*. Reprint, Reno: University of Nevada Press, 1991.

Schrader, Frank Charles. *Reconnaissance of the Jarbidge, Contact and Elko Mountain Mining Districts, Elko, Nevada*. Washington, D.C.: Government Printing Office, 1912.

Tuscarora Times-Review, 14 July 1897, January 28, 1902.

Falcon

REFERENCES

Elko County Recorder. *Miscellaneous Records*. Vols. 1, 2. 6 January 1878.

LaPointe, Daphne D., Joseph V. Tingley, and Richard B. Jones. *Mineral Resources of Elko County, Nevada*. Nevada Bureau of Mines Bulletin no. 106. Reno: Mackay School of Mines, University of Nevada, 1991.

Lincoln, Francis Church. *Mining Districts and Mineral Resources of Nevada*. Reno: Nevada Newsletter Publishing Company, 1923.

National Archives. Postal Records of Nevada. 1860–1970.

Tuscarora Times-Review, 14 November 1877, 24 April 1878, 2 November 1879, 12 August 1887.

Good Hope

REFERENCES

Angel, Myron, ed. *History of Nevada*. Oakland: Thompson and West, 1881. Reprint, Berkeley: Howell North Books, 1958.

Elko Independent, 18 July 1923.

LaPointe, Daphne D., Joseph V. Tingley, and Richard B. Jones. *Mineral Resources of Elko County, Nevada*. Nevada Bureau of Mines Bulletin no. 106. Reno: Mackay School of Mines, University of Nevada, 1991.

National Archives. Postal Records of Nevada. 1860–1970.

Tuscarora Times-Review, 24 January 1879, 30 May 1881, 22 September 1881, 6 April 1903.

Ivanhoe
REFERENCES

Lincoln, Francis Church. *Mining Districts and Mineral Resources of Nevada.* Reno: Nevada Newsletter Publishing Company, 1923.
Elko Free Press, 18 December 1920, 28 September 1940.
Elko Independent, 6 May 1930, 23 December 1939.
Smith, Roscoe. *Mineral Resources of Elko County.* USGS Open File Report no. 76–56. Washington, D.C.: Government Printing Office, 1976.

Jack Creek
REFERENCES

Elko Independent, 24 May 1877.
Elko Free Press, 24 June 1911.
National Archives. Postal Records of Nevada. 1860–1970.
Smith, Mary Urriola. "Memories of Jack Creek, 1925–1935." *Northeastern Nevada Historical Society Quarterly* 2 (1990): 30–44.
Tuscarora Times-Review, 16 November 1880, 24 April 1886, 6 May 1889, 28 January 1899, 16 June 1903.
U.S. Bureau of the Census. 1900.

NOTES

5. *Tuscarora Times-Review,* 16 November 1880.

Lime Mountain
REFERENCES

Couch, Betrand F., and Jay A. Carpenter. *Nevada's Metal and Mineral Production, 1859–1940.* University of Nevada Bulletin vol. 37, no. 4. Reno: University of Nevada, 19413.
Elko County Recorder. *Miscellaneous Records.* Vol. 1.
Elko Free Press, 24 March 1897, 4 May 1910, 16 April 1937.
Elko Independent, 12 April 1871.

Lone Mountain
REFERENCES

Couch, Betrand F., and Jay A. Carpenter. *Nevada's Metal and Mineral Production, 1859–1940.* University of Nevada Bulletin vol. 37, no. 4. Reno: University of Nevada, 1943.
Elko County Recorder. *Miscellaneous Records.* Vols. 1–3. 6 January 1888, 24 September 1880.
Elko Free Press, 12 August 1897, 2 August 1905, 14 June 1906, 24 September 1939.
Elko Independent, 9 November 1875, 3 June 1897, 14 May 1900, 8 February 1941.
Fairchild, Blawnie Mae. "Rip Van Winkle Mine." *Northeastern Nevada Historical Society Quarterly* 2 (spring 1985): 47–52.
Granger, Arthur E., Mendell M. Bell, George Simmons, and Florence Lee. *Geology*

and *Mineral Resources of Elko County, Nevada.* Nevada Bureau of Mines Bulletin no. 54. Reno: University of Nevada, 1957.

LaPointe, Daphne D., Joseph V. Tingley, and Richard B. Jones. *Mineral Resources of Elko County, Nevada.* Nevada Bureau of Mines Bulletin no. 106. Reno: Mackay School of Mines, University of Nevada, 1991.

Tuscarora Times-Review, 18 July 1881.

Midas

REFERENCES

Lincoln, Francis Church. *Mining Districts and Mineral Resources of Nevada.* Reno: Nevada Newsletter Publishing Company, 1923.

Couch, Betrand F., and Jay A. Carpenter. *Nevada's Metal and Mineral Production, 1859–1940.* University of Nevada Bulletin 37, no. 4. Reno: University of Nevada, 1943.

Elko Free Press, 12 March 1909, 17 April 1922, 4 August 1926, 16 June 1927, and various other issues.

Elko Independent, 10 April 1908, 24 April 1909, 24 April 1923, 18 February 1932.

Gold Circle News, various issues.

Humboldt Star, 4 October 1907, 6 November 1907.

Lingenfelter, Richard E., and Karen Rix Gash. *The Newspapers of Nevada: A History and Bibliography, 1854–1979.* Reno: University of Nevada Press, 1984.

National Archives. Postal Records of Nevada. 1860–1970.

Patterson, Edna, Louise Ulph, and Victor Goodwin. *Nevada's Northeast Frontier.* Reprint, Reno: University of Nevada Press, 1991.

Stevens, Sue Ann. "Midas, a Twentieth-Century Gold Camp." *Northeastern Nevada Historical Society Quarterly* 4 (fall 1984): 111–17.

Weed, Walter. *The Mines Handbook and Copper Handbook, 1916–1926.* Washington, D.C.: Government Printing Office, 1927.

Rio Tinto

REFERENCES

Angel, Myron, ed. *History of Nevada.* Oakland: Thompson and West, 1881. Reprint, Berkeley: Howell North Books, 1958.

Basanez, Dan. "Copper in the Copes: A History of Rio Tinto." *Northeastern Nevada Historical Society Quarterly* 4 (fall 1979): 95–107.

Davidson, P. W. "History of Early Jarbidge." N.d. Northeastern Nevada Museum manuscript collection, Elko, Nevada.

Elko Free Press, 14 August 1936.

Elko Independent, 2 July 1932.

Granger, Arthur E., Mendell M. Bell, and Florence Lee. *Geology and Mineral Resources of Elko County, Nevada.* Nevada Bureau of Mines Bulletin no. 54. Reno: University of Nevada, 1957.

National Archives. Postal Records of Nevada. 1860–1970.

Rio Tinto News, 18 January 1938.

Wells Progress, 16 November 1941.

Swales Mountain
REFERENCES

Elko Free Press, 12 June 1937.

Elko Independent, 27 March 1925.

Weed, Walter. *The Mines Handbook and Copper Handbook, 1916–1926.* Washington, D.C.: Government Printing Office, 1927.

Taylors
REFERENCES

Elko Independent, 6 April 1886.

Myrick, David F. *Railroads of Nevada and Eastern California: Volume 1, The Northern Roads.* Berkeley: Howell-North Books, 1962. Reprint. Reno: University of Nevada Press, 1992.

National Archives. Postal Records of Nevada. 1860–1970.

Patterson, Edna, Louise Ulph, and Victor Goodwin. *Nevada's Northeast Frontier.* Reprint, Reno: University of Nevada Press, 1991.

Tuscarora Times-Review, 13 August 1898.

Wells Progress, 16 January 1940.

Tuscarora
REFERENCES

Couch, Betrand F., and Jay A. Carpenter. *Nevada's Metal and Mineral Production, 1859–1940.* University of Nevada Bulletin 37, no. 4. Reno: University of Nevada, 1943.

Elko Free Press, 16 February 1931, 28 July 1906.

Elko Independent, 2 October 1873, May and June 1877.

Engineering and Mining Journal, July 29, 1876.

Granger, Arthur E., Mendell M. Bell, George C. Simmons, and Florence Lee. *Geology and Mineral Resources of Elko County, Nevada.* Nevada Bureau of Mines Bulletin no. 54. Reno: University of Nevada, 1957.

Knowles, Ardelle Plunkett. "Plunkett Family, Tuscarora, Nevada." *Northeastern Nevada Historical Society Quarterly* 3 (1989): 59–68.

Lingenfelter, Richard E., and Karen Rix Gash. *The Newspapers of Nevada: A History and Bibliography, 1854–1979.* Reno: University of Nevada Press, 1984.

Mining and Scientific Press, August 23, 1918.

National Archives. Postal Records of Nevada. 1860–1970.

Paher, Stanley. *Nevada Ghost Towns and Mining Camps.* Las Vegas: Nevada Publications, 1970.

Patterson, Edna, Louise Ulph, and Victor Goodwin. *Nevada's Northeast Frontier.* Reprint, Reno: University of Nevada Press, 1991.

Sewell, Harvey, and Ernest Clawson, "History of Tuscarora," 1916, Northeastern Nevada Museum manuscript collection, Elko, Nevada.

Townley, John. "Tuscarora." *Northeastern Nevada Historical Society Quarterly* 3 (summer/fall 1971): 6–37.

Tuscarora Mining Review, various issues.

Tuscarora Times-Review, 27 November 1880, 24 December 1880, 21 September 1883,

12 September 1877, various issues of 1880, 1881, 1887, 24 October 1889, 11 November 1889, 23 April 1892, 18 June 1897, various issues of 1897, 24 October 1898, 6 February 1900, 21 November 1903.

U.S. Census, 1870.

State Mineralogist of Nevada. Annual Report 1877.

Vanderburg, William O. *Placer Mining in Nevada.* University of Nevada Bulletin vol. 30, no. 4. Reno: University of Nevada, 1936.

NOTES

6. *Engineering and Mining Journal,* July 29, 1876.

7. *Tuscarora Times-Review,* 11 November 1889.

Warm Creek

REFERENCES

Morgan, Dale Lowell. *The Humboldt, High Road of the West.* New York: Farrar and Rinehart, 1943.

Elko Independent, 14 May 1870.

Elko County Recorder. Miscellaneous Records. Vol. 2. 23 June 1912, 8 December 1912.

LaPointe, Daphne D., Joseph V. Tingley, and Richard A. Jones. *Mineral Resources of Elko County, Nevada.* Nevada Bureau of Mines Bulletin no. 106. Reno: Mackay School of Mines, University of Nevada, 1991.

National Archives. Postal Records of Nevada. 1860–1970.

Nevada State Herald, 16 March 1912, 12 April 1922, 24 August 1953, 12 November 1941.

Patterson, Edna, Louise Ulph, and Victor Goodwin. *Nevada's Northeast Frontier.* Reprint, Reno: University of Nevada Press, 1991.

White Rock

REFERENCES

Garat, Margaret Brenner. "Family Archives: Garat and Indart, Brenner and Eastham." 1988. Northeastern Nevada Museum manuscript collection, Elko, Nevada.

Elko Independent, 23 March 1939, 12 April 1903.

National Archives. Postal Records of Nevada. 1860–1970.

Tuscarora Times-Review, various 1877 issues, various issues from October to December 1886, 18 March 1889.

North Central Elko County

Afton

REFERENCES

Bowen, Marshall E. "Bitter Times, the Summers of 1915 and 1916 on Northeast Nevada's Dry Farms." *Northeastern Nevada Historical Society Quarterly* 1 (1993): 2–21.

National Archives. Postal Records of Nevada. 1860–1970.

Nevada State Herald, 18 September 1916.

Alazon
REFERENCES

Nevada State Herald, 23 December 1948.

Alder
REFERENCES

LaPointe, Daphne D. Joseph V. Tingley, and Richard Jones. *Mineral Resources of Elko County, Nevada.* Nevada Bureau of Mines Bulletin no. 106. Reno: Mackay School of Mines, University of Nevada, 1991.
Nevada State Herald, 23 December 1948.

Bruno City
REFERENCES

Elko Daily Free Press, 14 January 1903.
Elko Independent, 18 July 1875, 19 June 1876.
LaPointe, Daphne D., Joseph V. Tingley, and Richard B. Jones. *Mineral Resources of Elko County, Nevada.* Nevada Bureau of Mines Bulletin no. 106. Reno: Mackay School of Mines, University of Nevada, 1991.

Charleston
REFERENCES

Elko County Recorder. *Mining Locations.* Vol. 1.
Elko County Recorder. *Miscellaneous Records.* Vol. 1.
Elko Daily Free Press, 15 April 1916.
Elko Independent, 8 August 1911.
Hickson, Howard. "Murder, Better than Fingerprints." *Northeastern Nevada Historical Society Quarterly* 1 (winter 1985): 13–17.
National Archives. Postal Records of Nevada. 1860–1970.
Nevada State Herald, 23 June 1901.
Patterson, Edna, Louise Ulph, and Victor Goodwin. *Nevada's Northeast Frontier.* Reprint, Reno: University of Nevada Press, 1991.
Tuscarora Times-Review, 13 September 1886, 14 July 1889, 12 June 1897, 25 June 1900.

NOTES

1. Edna Patterson, Louise Ulph, and Victor Goodwin, *Nevada's Northeast Frontier* (Reprint, Reno: University of Nevada Press, 1991).

Coal Canyon
REFERENCES

Elko Independent, 2 October 1936.
LaPointe, Daphne D., Joseph V. Tingley, and Richard B. Jones. *Mineral Resources of Elko County, Nevada.* Nevada Bureau of Mines Bulletin no. 106. Reno: Mackay School of Mines, University of Nevada, 1991.

Elk Mountain

REFERENCES

Elko County Recorder. *Miscellaneous Records.* Vol. 3.
Schrader, Frank Charles. *Reconnaissance of the Jarbidge, Contact and Elk Mountain Mining Districts, Elko County, Nevada.* Washington, D.C.: Government Printing Office, 1912.

Gold Creek

REFERENCES

Elko County Recorder. *Miscellaneous Records.* Vol. 1.
Elko Free Press, 28 August 1897, 23 July 1898.
Elko Independent, 19 October 1873.
Gold Creek News, 20 February 1897, 12 May 1897.
Lingenfelter, Richard E., and Karen Rix Gash. *The Newspapers of Nevada: A History and Bibliography, 1854–1979.* Reno: University of Nevada Press, 1984.
National Archives. Postal Records of Nevada. 1860–1970.
Tuscarora Times-Review, 12 July 1886, 3 July 1897, 6 May 1898, 24 October 1899, 28 January 1900, 12 January 1929.
U.S. Bureau of the Census. 1880.

Hicks District

REFERENCES

Elko County Recorder. *Miscellaneous Records.* Vols. 1, 2.
———. *Mining Locations.* Vol. 1.
Elko Free Press, 24 July 1915.
Elko Independent, 14 August 1876.
LaPointe, Daphne D., Joseph V. Tingley, and Richard E. Jones. *Mineral Resources of Elko County, Nevada.* Nevada Bureau of Mines Bulletin no. 106. Reno: Maackay School of Mines, University of Nevada, 1991.
Hicks to Dearing, 30 November 1882.
U.S. Bureau of the Census. 1880.

NOTES

2. Hicks to Dearing, 30 November 1882.

Hubbard

REFERENCES

Wells Progress, 27 August 1938.

Ivada

REFERENCES

Elko Free Press, 14 May 1910, 2 July 1965.

NOTES

3. *Elko Free Press,* 2 July 1965.

Jarbidge
REFERENCES

Copeland, Teresa. "Jarbidge." N.d. Northeastern Nevada Museum manuscript collection, Elko, Nevada.

Couch, Betrand F. and Jay A. Carpenter. *Nevada's Metal and Mineral Production, 1859–1940.* University of Nevada Bulletin 37, no. 4. Reno: University of Nevada, 1943.

Elko Independent, 16 July 1917, 16 September 1933.

Hickson, Howard. "Last Horsedrawn Stage Robbery." *Northeastern Nevada Historical Society Quarterly* 1 (winter 1981): 2–30.

LaPointe, Daphne D. Joseph V. Tingley, and Richard B. Jones. *Mineral Resources of Elko County, Nevada.* Nevada Bureau of Mines Bulletin no. 106. Reno: Mackay School of Mines, University of Nevada, 1991.

Lincoln, Francis Church. *Mining Districts and Mineral Resources of Nevada.* Reno: Nevada Newsletter Publishing Company, 1923.

National Archives. Postal Records of Nevada. 1860–1970.

Northeastern Nevada Museum research files.

Patterson, Edna, Louise Ulph, and Victor Goodwin. *Nevada's Northeast Frontier.* Reprint, Reno: University of Nevada Press, 1991.

Schrader, Frank Charles. *Reconnaissance of the Jarbidge, Contact and Elk Mountain Mining Districts, Elko County, Nevada.* Washington, D.C.: Government Printing Office, 1912.

Wilson, Helen E. *Gold Fever.* La Mesa, Calif.: Helen E. Wilson, 1974.

NOTES

4. *Elko Independent,* 16 June 1917.

Metropolis
REFERENCES

Bowen, Marshall. "Bitter Times: The Summers of 1915 and 1916 on Northeast Nevada's Dry Farms." *Northeastern Nevada Historical Society Quarterly* 1 (1993): 2–21.

Halton, Stephanie. "Metropolis." N.d. Northeastern Nevada Museum research files, Elko, Nevada.

Holbrook, Marjorie. *History of Metropolis, Nevada.* Marjorie Holbrook, 1986.

Lingenfelter, Richard E., and Karen Rix Gash. *The Newspapers of Nevada: A History and Bibliography, 1854–1979.* Reno: University of Nevada Press, 1984.

National Archives. Postal Records of Nevada. 1860–1970.

Nevada State Herald, 23 July 1915, 12 October 1915, 25 February 1916, 8 June 1925, 30 September 1929, 12 September 1936.

Woelz, William. "Metropolis: Death of a Dream." *Northeastern Nevada Historical Society Quarterly* 1 (spring 1973): 3–14.

Mountain City
REFERENCES

Elko Independent, 29 May 1869, 18 July 1869, 8 July 1870, 12 November 1871.

Frampton, Fred. "Excavation of Two Chinese Dugouts in Placerville, Nevada." 1994. U.S. Forest Service files, Elko, Nevada.

Granger, Arthur E., Mendell M. Bell, George C. Simmons, and Florence Lee. *Geology and Mineral Resources of Elko County, Nevada*. Nevada Bureau of Mines and Geology Bulletin no. 54. Reno: University of Nevada, 1957.

LaPointe, Daphne D., Joseph V. Tingley, and Richard B. Jones. *Mineral Resources of Elko County, Nevada*. Nevada Bureau of Mines Bulletin no. 106. Reno: Mackay School of Mines, University of Nevada, 1991.

Lingenfelter, Richard E., and Karen Rix Gash. *The Newspapers of Nevada: A History and Bibliography, 1854–1979*. Reno: University of Nevada Press, 1984.

Mountain City Times, 28 January 1898.

National Archives. Postal Records of Nevada. 1860–1970.

Paher, Stanley. *Nevada Ghost Towns and Mining Camps*. Las Vegas: Nevada Publications, 1970.

Patterson, Edna, Louise Ulph, and Victor Goodwin. *Nevada's Northeast Frontier*. Reprint, Reno: University of Nevada Press, 1991.

State Mineralogist of Nevada, Annual Report, 1869.

Tuscarora Times-Review, 25 May 1887, 6 June 1888, 22 December 1889.

U.S. Bureau of the Census. 1900.

U.S. Forest Service files. Elko, Nevada.

North Fork
REFERENCES

Aitchison, Pat Morse. *Morse Family Treasures*. Salt Lake City: Circulation Service, Inc., 1990.

Elko Free Press, 20 March 1987.

Elko Independent, 6 August 1934.

National Archives. Postal Records of Nevada. 1860–1970.

Patterson, Edna, Louise Ulph, and Victor Goodwin. *Nevada's Northeast Frontier*. Reprint, Reno: University of Nevada Press, 1991.

Reed, Flo. *Bygone Days of Nevada Schools*. Flo Reed, 1991.

Robertson, Opal Curieux. "William and Catherine Johnson Family, North Fork, Pioneers." *Northeastern Nevada Historical Society Quarterly* 2 (1991): 51–74.

U.S. Bureau of the Census. 1880 and 1900.

Patsville
REFERENCES

Elko Independent, 12 May 1937.

Patterson, Edna. *Who Named It?* Elko: Elko Independent Publishing, 1965.

Rowland
REFERENCES

Aalseth, Ruthe Scott. "Memories of Rowland, Nevada." *Northeastern Nevada Historical Society Quarterly* 4 (1995): 176–81.

Barton, Mary Scott. *The Life of the John B. Scott Family*. Mary Scott Barton, 1986.

Elko Free Press, 23 August 1927.

Elko Independent, 15 September 1925.

LaPointe, Daphne D., Joseph V. Tingley, and Richard B. Jones. *Mineral Resources of Elko County, Nevada,* Nevada Bureau of Mines Bulletin #106: Reno: Mackay School of Mines, University of Nevada, 1991.

National Archives. Postal Records of Nevada. 1860–1970.

Patterson, Edna, Louise Ulph, and Victor Goodwin. *Nevada's Northeast Frontier.* Reprint, Reno: University of Nevada Press, 1991.

Strickland, Ed. Northeastern Nevada Museum oral history files, Elko, Nevada.

Stofiel
REFERENCES

Elko Independent, 12 May 1905.

National Archives. Postal Records of Nevada. 1860–1970.

Telephone Mining District
REFERENCES

Tuscarora Times-Review, 25 August 1885.

Tulasco
REFERENCES

Wells Progress, 12 June 1936.

Northeastern Elko County

Alabama Mining District
REFERENCES

LaPointe, Daphne D., Joseph V. Tingley, and Richard B. Jones. *Mineral Resources of Elko County, Nevada.* Nevada Bureau of Mines and Geology Bulletin no. 106. Reno: Mackay School of Mines, University of Nevada, 1991.

Patterson, Edna. "What's in a Name?" *Northeastern Nevada Historical Society Quarterly* 2 (spring/summer 1977): 31–45.

Annaville
REFERENCES

LaPointe, Daphne D., Joseph V. Tingley, and Richard B. Jones. *Mineral Resources of Elko County, Nevada.* Nevada Bureau of Mines and Geology Bulletin no. 106. Reno: Mackay School of Mines, University of Nevada, 1991.

Buel
REFERENCES

Couch, Bertrand F., and Jay Carpenter. *Nevada's Metal and Mineral Production, 1859–1940.* University of Nevada Bulletin vol. 37, no. 4. Reno: University of Nevada, 1943.

Elko County Recorder. *Miscellaneous Records.* Vol. 1.

LaPointe, Daphne D., Joseph V. Tingley, and Richard B. Jones. *Mineral Resources of*

Elko County, Nevada. Nevada Bureau of Mines and Geology Bulletin no. 106. Reno: Mackay School of Mines, University of Nevada, 1991.

Myrick, David. *Railroads of Nevada and Eastern California: Volume 1, The Northern Roads.* Berkeley: Howell-North, 1963. Reprint. Reno: University of Nevada Press, 1992.

National Archives. Postal Records of Nevada. 1860–1970.

Paher, Stanley. *Nevada Ghost Towns and Mining Camps.* Las Vegas: Nevada Publications, 1970.

Cobre
REFERENCES

National Archives. Postal Records of Nevada. 1860–1970.
Nevada State Herald, 14 June 1906, 30 August 1907, 3 April 1908, 21 December 1921.
Wells Progress, 30 November 1936, 14 May 1938.

NOTES

1. *Wells Progress,* 30 November 1936.

Contact
REFERENCES

Carlson, Helen. *Nevada Place Names.* Reno: University of Nevada, 1974.

Elko County Recorder. *Miscellaneous Records.* Vol. 1.

Elko Free Press, 12 February 1908, 2 April 1909.

Patterson, Marguerite Evans. "Letters from Contact." *Northeastern Nevada Historical Society Quarterly* 1 (winter 1988): 3–14.

Hickson, Howard. "Where's the Beef?" *Northeastern Nevada Historical Society Quarterly* 3 (1990): 81–83.

Lincoln, Francis Church. *Mining Districts and Mineral Resources of Nevada.* Nevada Newsletter Publishing Company, 1923.

Lingenfelter, Richard E., and Karen Rix Gash. *The Newspapers of Nevada: A History & Bibliography, 1854–1979.* Reno: University of Nevada Press, 1984.

National Archives. Postal Records of Nevada. 1860–1970.

Nevada State Herald, 30 September 1922, 26 May 1924, 27 December 1924, 14 June 1936, 5 August 1942.

Patterson, Edna, Louise Ulph, and Victor Goodwin. *Nevada's Northeast Frontier.* Reprint, Reno: University of Nevada Press, 1991.

Reed, Flo. *Bygone Days of Nevada Schools.* Flo Reed, 1991.

Schrader, Frank Charles. *Reconnaissance of the Jarbidge, Contact and Elk Mountain Mining Districts, Elko County, Nevada.* Washington, D.C.: Government Printing Office, 1912.

Smith, Roscoe. *Mineral Resources of Elko County.* USGS Open File Report no. 76–56. Washington, D.C.: Government Printing Office, 1976.

NOTES

2. *Elko Free Press,* 2 April 1909.

3. Marguerite Patterson Evans, "Letters from Contact," *Northeastern Nevada Historical Society Quarterly* 1 (1988): 3–14.

4. Ibid.

Delano
REFERENCES

Angel, Myron ed. *History of Nevada.* Oakland: Thompson and West, 1881. Reprint, Berkeley: Howell North Books, 1958.

Couch, Bertrand F., and Jay Carpenter. *Nevada's Metal and Mineral Production, 1859–1940.* University of Nevada Bulletin vol. 37, no. 4. Reno: University of Nevada, 1943.

Elko County Recorder. *Mining Locations.* Vol. 1.

Elko County Recorder. *Miscellaneous Records.* Vol. 3.

Elko Independent, 14 May 1875, 5 September 1884, 24 August 1892.

LaPointe, Daphne D., Joseph V. Tingley, and Richard B. Jones. *Mineral Resources of Elko County, Nevada.* Nevada Bureau of Mines and Geology Bulletin no. 106. Reno: Mackay School of Mines, University of Nevada, 1991.

National Archives. Postal Records of Nevada. 1860–1970.

Nevada State Herald, 27 September 1927, 6 May 1936.

Loray
REFERENCES

Elko County Recorder. *Miscellaneous Records.* Vol. 2.

Elko Independent, 14 May 1875.

Granger, Arthur E., Mendell M. Bell, George C. Simmons, and Florence Lee. *Geology and Mineral Resources of Elko County, Nevada.* Nevada Bureau of Mines Bulletin no. 54. Reno: University of Nevada, 1957.

Shearer, Frederick. *Pacific Tourist.* New York: Adams & Bishop, 1881.

Montello
REFERENCES

Earl, Phillip. "The Montello Robbery." *Northeastern Nevada Historical Society Quarterly* 2 (summer 1972): 3–17.

Elko Free Press, 12 March 1883.

National Archives. Postal Records of Nevada. 1860–1970.

Nevada State Herald, 30 October 1925.

Patterson, Edna, Louise Ulph, and Victor Goodwin. *Nevada's Northeast Frontier.* Reprint, Reno: University of Nevada Press, 1991.

Wells Progress, 15 March 1936.

Pilot
REFERENCES

Elko Independent, 7 August 1878.

LaPointe, Daphne D., Joseph V. Tingley, and Richard B. Jones. *Mineral Resources of*

Elko County, Nevada. Nevada Bureau of Mines and Geology Bulletin no. 106. Reno: Mackay School of Mines, University of Nevada, 1991.

Wells Progress, 15 March 1936.

Toano
REFERENCES

Elko Independent, 12 November 1871, 23 December 1871, 11 February 1873, 20 June 1874, 12 June 1881.

National Archives. Postal Records of Nevada. 1860–1970.

Nevada State Herald, November 16, 1936.

Patterson, Edna, Louise Ulph, and Victor Goodwin. *Nevada's Northeast Frontier.* Reprint, Reno: University of Nevada Press, 1991.

Shearer, Frederick. *Pacific Tourist.* New York: Adams & Bishop, 1881.

U.S. Bureau of the Census. 1870.

Wilkins
REFERENCES

Bowman, Nora Linjer. *Only the Mountains Remain.* Caldwell, Idaho: Caxton Printers, 1958.

Elko County Recorder. *Mining Locations.* Vol. 1.

Moschetti, John. Oral history. 1995. Northeastern Nevada Museum oral history files, Elko, Nevada.

Nevada State Herald, 16 February 1911.

Wells Progress, 14 August 1953.

Southwestern Elko County

Arthur
REFERENCES

Elko Free Press, 6 November 1899, 15 November 1899.

National Archives. Postal Records of Nevada, 1860–1970.

Patterson, Edna, Louise Ulph, and Victor Goodwin. *Nevada's Northeast Frontier.* Reprint, Reno: University of Nevada Press, 1991.

Smales, Leonard. "Arthur Gedney: Pioneer Homesteader." N.d. Northeastern Nevada Museum manuscript collection, Elko, Nevada.

NOTES

1. *Elko Free Press,* 15 November 1899.
2. Ruhlen, George, *Early Nevada Forts* (Reno: Nevada Historical Society, 1964).

Battle Creek
REFERENCES

LaPointe, Daphne D. Joseph V. Tingley, and Richard B. Jones. *Mineral Resources of Elko County.* Nevada Bureau of Mines Bulletin no. 106. Reno: Mackay School of Mines, University of Nevada, 1991.

National Archives. Postal Records of Nevada. 1860–1970.

Patterson, Edna. "What's in a Name?" *Northeastern Nevada Historical Society Quarterly* 2 (spring/summer 1977): 3–45.

Patterson, Edna, Louise Ulph, and Victor Goodwin. *Nevada's Northeast Frontier*. Reprint, Reno: University of Nevada Press, 1991.

Wells Progress, 18 October 1951.

Bullion
REFERENCES

Angel, Myron, ed. *History of Nevada*. Oakland: Thompson and West, 1881. Reprint, Berkeley: Howell North Books, 1958.

Couch, Betrand F., and Jay Carpenter. *Nevada's Metal and Mineral Production, 1859–1940*. University of Nevada Bulletin 37, no. 4. Reno: University of Nevada, 1943.

Elko Independent, 18 August 1872, 13 October 1876, 12 September 1883, 8 February 1888.

Emmons, William. *A Reconnaissance of Some Mining Camps in Elko, Lander, and Eureka Counties, Nevada*. USGS Bulletin no. 408. Washington, D.C.: Government Printing Office, 1910.

LaPointe, Daphne D., Joseph V. Tingley, and Richard Jones. *Mineral Resources of Elko County, Nevada*. Nevada Bureau of Mines Bulletin no. 106. Reno: Mackay School of Mines, University of Nevada, 1991.

National Archives. Postal Records of Nevada. 1860–1970.

Patterson, Edna, Louise Ulph, and Victor Goodwin, *Nevada's Northeast Frontier*. Reprint, Reno: University of Nevada Press, 1991.

Reed, Flo. *Bygone Days of Nevada Schools*. Flo Reed, 1991.

Tuscarora Times-Review, July 18, 1886.

Cave Creek
REFERENCES

Angel, Myron, ed. *History of Nevada*. Oakland: Thompson and West, 1881. Reprint, Berkeley: Howell North Books, 1958.

Davis, James. "Harry Gallagher Fish Hatchery." *Northeastern Nevada Historical Society Quarterly* 1 (winter 1976): 3–13.

National Archives. Postal Records of Nevada. 1860–1970.

Patterson, Edna, Louise Ulph, and Victor Goodwin. *Nevada's Northeast Frontier*. Reprint, Reno: University of Nevada Press, 1991.

Deeth
REFERENCES

Elko Free Press, 10 November 1956.

Elko Independent, 14 January 1915, 20 October 1915.

Lingenfelter, Richard E., and Karen Rix Gash. *The Newspapers of Nevada: A History & Bibliography, 1854–1979*. Reno: University of Nevada Press, 1984.

Nevada State Herald, 18 September 1903, February 1, 1915.

OCTA Newsletter, January 1991.

Patterson, Edna, Louise Ulph, and Victor Goodwin. *Nevada's Northeast Frontier.* Reprint, Reno: University of Nevada Press, 1991.

Delker
REFERENCES

Hill, James. *Notes on some Mining Districts in Eastern Nevada.* USGS Bulletin no. 648. Washington, D.C.: Government Printing Office, 1916.

LaPointe, Daphne D., Joseph V. Tingley, and Richard B. Jones. *Mineral Resources of Elko County, Nevada.* Nevada Bureau of Mines Bulletin no. 106. Reno: Mackay School of Mines, University of Nevada, 1991.

Nevada State Herald, 23 August 1907.

Smith, Roscoe. *Mineral Resources of Elko County.* USGS Open File Report no. 76–56. Washington, D.C.: Government Printing Office, 1976.

Fort Halleck
REFERENCES

Elko Independent, 15 October 1870, 22 April 1897.

National Archives. Postal Records of Nevada. 1860–1970.

Patterson, Edna, Louise Ulph, and Victor Goodwin. *Nevada's Northeast Frontier.* Reprint, Reno: University of Nevada Press, 1991.

Ruhlen, George. *Early Nevada Forts.* Reno: Nevada Historical Society, 1964.

Halleck
REFERENCES

DeVries, Jennifer. "History of the 71 Ranch." N.d. Northeastern Nevada Museum manuscript collection, Elko, Nevada.

National Archives. Postal Records of Nevada. 1860–1970.

Patterson, Edna, Louise Ulph, and Victor Goodwin. *Nevada's Northeast Frontier.* Reprint, Reno: University of Nevada Press, 1991.

Patterson, Edna. "What's In a Name?" *Northeastern Nevada Historical Society Quarterly* 2 (spring/summer 1977): 3–45.

Harrison Pass
REFERENCES

Angel, Myron, ed. *History of Nevada.* Oakland: Thompson and West, 1881. Reprint, Berkeley: Howell North Books, 1958.

Elko Independent, 23 January 1897.

LaPointe, Daphne D., Joseph V. Tingley, and Richard B. Jones. *Mineral Resources of Elko County, Nevada.* Nevada Bureau of Mines Bulletin no. 106. Reno: Mackay School of Mines, University of Nevada, 1991.

Wells Progress, 12 April 1941.

Jiggs

REFERENCES

Beasley, Teri. "J. J. Hylton, Jiggs Area Pioneer." *Northeastern Nevada Historical Society Quarterly* 4 (1990): 91–99.

Hickson, Howard. "Jiggs and Mound Valley." *Northeastern Nevada Historical Society Quarterly* 3 (summer 1988): 59–71.

McCartney, Mary. "Maggie Club File." 1995. Northeastern Nevada Museum manuscript collection, Elko, Nevada.

National Archives. Postal Records of Nevada. 1860–1970.

Patterson, Edna, Louise Ulph, and Victor Goodwin. *Nevada's Northeast Frontier.* Reprint, Reno: University of Nevada Press, 1991.

Porch family history. Maggie Club files. 1995. Northeastern Nevada Museum manuscript collection, Elko, Nevada.

U.S. Bureau of the Census. 1900

Lamoille

REFERENCES

Buzetti, Mitch. "Elko-Lamoille Power Company." *Northeastern Nevada Historical Society Quarterly* 4 (fall 1987): 81–87.

Elko County School Records.

Elko Free Press, 12 October 1915.

Elko Independent, 27 June 1869.

National Archives. Postal Records of Nevada. 1860–1970.

Northeastern Nevada Museum research files.

Patterson, Edna. *This Land Was Ours.* Springville, Utah: Art City Publishing Co., 1973.

———. "Little Church of the Crossroads." *Northeastern Nevada Historical Society Quarterly* 3 (summer 1986): 67–74.

Patterson, Edna, Louise Ulph, and Victor Goodwin. *Nevada's Northeast Frontier.* Reprint, Reno: University of Nevada Press, 1991.

U.S. Bureau of the Census. 1880

Larrabee Mining District

REFERENCES

Elko County Recorder. *Miscellaneous Records.* Vol. 2.

LaPointe, Daphne D., Joseph V. Tingley, and Richard B. Jones. *Mineral Resources of Elko County, Nevada.* Nevada Bureau of Mines Bulletin no. 106. Reno: Mackay School of Mines, University of Nevada, 1991.

Lee

REFERENCES

Carlson, Helen. *Nevada Place Names.* Reno: University of Nevada Press, 1974.

Elko Free Press, 24 September 1917.

Elko Independent, 19 November 1881, 26 July 1925, 18 October 1927.

LaPointe, Daphne D., Joseph V. Tingley, and Richard B. Jones. *Mineral Resources of*

Elko County, Nevada. Nevada Bureau of Mines Bulletin no. 106. Reno: Mackay School of Mines, University of Nevada, 1991.

Patterson, Edna, Louise Ulph, and Victor Goodwin. *Nevada's Northeast Frontier.* Reprint, Reno: University of Nevada Press, 1991.

Reed, Flo. *Bygone Days of Nevada Schools.* Flo Reed, 1991.

Weed, Walter. *The Copper Handbook.* Washington, D.C.: Government Printing Office, 1912.

Mud Springs
REFERENCES

Elko Independent, 6 March 1936.

Hill, James M. *Notes on Some Mining Districts in Eastern Nevada.* USGS Bulletin no. 648. Washington, D.C.: Goverment Printing Office, 1916.

LaPointe, Daphne D., Joseph V. Tingley, and Richard B. Jones. *Mineral Resources of Elko County, Nevada.* Nevada Bureau of Mines Bulletin no. 106. Reno: Mackay School of Mines, University of Nevada, 1991.

Lincoln, Francis Church. *Mining Districts and Mineral Resources of Nevada.* Reno: Nevada Newsletter Publishing Company, 1923.

Nevada State Herald, 23 August 1924, 17 September 1934.

Wells Progress, 25 August 1951.

Racine Mining District
REFERENCES

Elko County Recorder. *Miscellaneous Records.* Vol. 1.

Elko Independent, 25 September 1872, 12 July 1874.

Ruby City
REFERENCES

National Archives. Postal Records of Nevada. 1860–1970.

Palacio, Robert. "Ruby City." N.d. Northeastern Nevada Museum manuscript collection, Elko, Nevada.

Patterson, Edna, Louise Ulph, and Victor Goodwin. *Nevada's Northeast Frontier.* Reprint, Reno: University of Nevada Press, 1991.

Sherman
REFERENCES

Botsford, Louise Walther. "Valentine Walter, Nevada Pioneer." *Northeastern Nevada Historical Society Quarterly* 1 (1989): 3–10.

Elko Independent, May 11, 1886.

National Archives. Postal Records of Nevada. 1860–1970.

South Fork
REFERENCES

Elko Chronicle, advertisements appearing during 1870.

National Archives. Postal Records of Nevada. 1860–1970.

Patterson, Edna, Louise Ulph, and Victor Goodwin. *Nevada's Northeast Frontier.* Reprint, Reno: University of Nevada Press, 1991.

Valley View
REFERENCES

Elko Independent, 25 October 1927.

LaPointe, Daphne D., Joseph V. Tingley, and Richard B. Jones. *Mineral Resources of Elko County, Nevada.* Nevada Bureau of Mines Bulletin no 106. Reno: Mackay School of Mines, University of Nevada, 1991.

Wells Progress, 13 February 1954.

Southeastern Elko County

Currie
REFERENCES

Elko Daily Free Press, 14 July 1907.

Friends of the Nevada Northern Railway, *Nevada Northern Railway and the Copper Camps of White Pine County, Nevada.* East Ely, Nev.: Friends of the Nevada Northern Railway, n.d.

National Archives. Postal Records of Nevada. 1860–1970.

Reynolds, Leona. "Memories of Currie, Nevada." *Northeastern Nevada Historical Society Quarterly* 1 (spring 1975): 3–23.

Wells Progress, 23 November 1939.

Decoy
REFERENCES

Bowen, Marshall. "Bitter Times: The Summers of 1915 and 1916 on Northeast Nevada's Dry Farms." *Northeastern Nevada Historical Society Quarterly* 1 (1993): 2–21.

LaPointe, Daphne D., Joseph V. Tingley, and Richard B. Jones *Mineral Resources of Elko County, Nevada.* Nevada Bureau of Mines Bulletin no. 106. Reno: Mackay School of Mines, University of Nevada, 1991.

Dolly Varden
REFERENCES

Carlson, Helen. *Nevada Place Names.* Reno: University of Nevada Press, 1974.

Couch, Bertrand F., and Jay Carpenter. *Nevada's Metal and Mineral Production, 1859–1940.* University of Nevada Bulletin vol. 37, no. 4. Reno: University of Nevada, 1943.

Hill, James. *Notes on some Mining Districts in Eastern Nevada.* Washington: Government Printing Office, 1912.

LaPointe, Daphne D., Joseph V. Tingley, and Richard B. Jones. *Mineral Resources of Elko County, Nevada.* Nevada Bureau of Mines Bulletin no. 106. Reno: Mackay School of Mines, University of Nevada, 1991.

National Archives. Postal Records of Nevada. 1860–1970.

State Mineralogist of Nevada. Annual Report. 1875.

Wells Progress, 14 October 1953.

U.S. Bureau of the Census. 1880.

Ferber

REFERENCES

Elko County Recorder. *Mining Locations*. Vol. 2.

Elko County Recorder. *Miscellaneous Records*. Vol. 2.

Gianella, Vincent. *Nevada's Common Minerals*. University of Nevada Bulletin no. 35. Reno: University of Nevada, 1939.

LaPointe, Daphne D., Joseph V. Tingley, and Richard B. Jones. *Mineral Resources of Elko County, Nevada*. Nevada Bureau of Mines Bulletin no. 106. Reno: Mackay School of Mines, University of Nevada, 1991.

Ferguson Springs

REFERENCES

Hill, James. *Notes on some Mining Districts in Eastern Nevada*. Washington, D.C.: Government Printing Office, 1912.

LaPointe, Daphne D., Joseph V. Tingley, John, and Richard B. *Mineral Resources of Elko County, Nevada*. Nevada Bureau of Mines Bulletin no. 106. Reno: Mackay School of Mines, University of Nevada, 1991.

Tuscarora Times-Review, 24 June 1894.

Wells Progress, 14 June 1937.

Kingsley

REFERENCES

Browne, J. R. *Reports on Mineral Resources of the States and Territories West of the Rocky Mountains*. Washington, D.C.: Government Printing Office, 1867.

Elko County Recorder. *Mining Locations*. Vol. 1.

Elko Free Press, 4 February 1995.

Elko Independent, 16 July 1881, 28 April 1883.

Hickson, Howard. "Jiggs and Mound Valley." *Northeastern Nevada Historical Society Quarterly* 3 (summer 1988): 58–73.

Hill, James. *Notes on some Mining Districts in Eastern Nevada*. Washington, D.C.: Government Printing Office, 1912.

LaPointe, Daphne D., Joseph V. Tingley, and Richard B. Jones. *Mineral Resources of Elko County, Nevada*. Nevada Bureau of Mines Bulletin no. 106. Reno: Mackay School of Mines, University of Nevada, 1991.

Nevada State Herald, 28 December 1955.

Tuscarora Times-Review, 19 December 1886.

Lafayette Mining District

REFERENCES

LaPointe, Daphne D., Joseph V. Tingley, and Richard B. Jones. *Mineral Resources of Elko County, Nevada*. Nevada Bureau of Mines Bulletin no. 106. Reno: Maackay School of Mines, University of Nevada, 1991.

Proctor

REFERENCES

Nevada State Herald, 4 March 1910.

U.S. Geological Survey. *Mineral Resources of the United States, 1921.* Washington: Government Printing Office, 1921.

Shafter

REFERENCES

National Archives. Postal Records of Nevada. 1860–1970.

Patterson, "What's in a Name?" *Northeastern Nevada Historical Society Quarterly* 2 (spring/summer 1977): 3–45.

Wells Progress, 12 January 1941, 16 February 1950.

NOTES

1. *Nevada State Herald,* 6 March 1950.

Silver Zone

REFERENCES

Elko County Recorder. *Miscellaneous Records.* Vol. 2.

LaPointe, Daphne D., Joseph V. Tingley, and Richard B. Jones. *Mineral Resources of Elko County, Nevada.* Nevada Bureau of Mines Bulletin no. 106. Reno: Mackay School of Mines, University of Nevada, 1991.

National Archives. Postal Records of Nevada. 1860–1970.

Patterson, "What's in a Name?" *Northeastern Nevada Historical Society Quarterly* 2 (spring/summer 1977): 3–45

State Mineralogist of Nevada. Annual Report. 1873.

Wells Progress, 12 March 1936, 15 July 1942.

Spruce Mountain

REFERENCES

Couch, Bertrand F., and Jay Carpenter. *Nevada's Metal and Mineral Production, 1859–1940.* University of Nevada Bulletin 37, no. 4. Reno: University of Nevada, 1943.

Elko Free Press, 7 June 1885.

Elko Independent, August 14, 1871, 3 October 1871, 27 April 1880, 16 July 1883.

LaPointe, Daphne D., Joseph V. Tingley, and Richard B. Jones. *Mineral Resources of Elko County, Nevada.* Nevada Bureau of Mines Bulletin no. 106. Reno: Mackay School of Mines, University of Nevada, 1991.

National Archives. Postal Records of Nevada. 1860–1970.

Nevada State Herald, 30 March 1901, 23 April 1901, 6 March 1905, 20 February 1907, 23 August 1907, 2 March 1918, 27 January 1920, 6 August 1924, 2 May 1926, 27 May 1927.

Paher, Stanley. *Nevada Ghost Towns and Mining Camps.* Las Vegas: Nevada Publications, 1970.

Schrader, F. C. *Spruce Mountain District, Elko County, Nevada.* Nevada Bureau of Mines and Geology Bulletin no. 14. Reno: University of Nevada, 1931.

Smith, Roscoe. *Mineral Resources of Elko County.* USGS Open File Report no. 76–56. Washington, D.C.: Government Printing Office, 1976.

Tuscarora Times-Review, 24 January 1886.

Venable, B.W. *The Wisemans.* B.W. Venable, 1992.

Wells Progress, 31 May 1932, 12 July 1935, 12 November 1940, 6 September 1952, 6 January 1959.

Tobar

REFERENCES

Bowen, Marshall. "Dryland Homesteading on Tobar Flat," *Northeastern Nevada Historical Society Quarterly* 4 (fall 1981): 119–37.

Elko Free Press, 20 June 1969.

Lingenfelter, Richard E., and Karen Rix Gash. *The Newspapers of Nevada: A History & Bibliography, 1854–1979.* Reno: University of Nevada Press, 1984.

National Archives. Postal Records of Nevada. 1860–1970.

Nevada State Herald, 20 September 1923, 6 August 1942.

Paher, Stanley. *Nevada Ghost Towns and Mining Camps.* Las Vegas: Nevada Publications, 1973.

Powell, Charles Stewart. "Depression Days in Tobar." *Northeastern Nevada Historical Society Quarterly* 4 (fall 1978): 127–40.

Reese River Reveille, 12 June 1912.

Tatomer, William. "Tobar." *Northeastern Nevada Historical Society Quarterly* 1 (winter 1978): 3–11.

Valley Pass

REFERENCES

Wells Progress, 16 July 1939.

Ventosa

REFERENCES

Wells Progress, 31 August 1950.

Wardell

REFERENCES

Elko Independent, 28 October 1880.

Nevada State Herald, 12 July 1905.

White Horse

REFERENCES

Carlson, Helen. *Nevada Place Names.* Reno: University of Nevada Press, 1974.

LaPointe, Daphne D., Joseph V. Tingley, and Richard B. Jones. *Mineral Resources of Elko County, Nevada.* Nevada Bureau of Mines Bulletin no. 106. Reno: Mackay School of Mines, University of Nevada, 1991.

Selected Bibliography

Books and Articles

Aitchison, Pat Morse. *Morse Family Treasures*. Salt Lake City: Circulation Service, Inc., 1990.

Anderson, Ruth. *Memoirs of Leona Reynolds*. Carson City, Nev.: Bicentennial Commission, 1976.

Angel, Myron, ed. *History of Nevada*. Oakland: Thompson and West, 1881. Reprint, Berkeley: Howell North Books, 1958.

Armstrong, Robert D. *A Preliminary Union Catalog of Nevada Manuscripts*. Reno: University of Nevada Library Association, 1967.

————. *Nevada Printing History, A Bibliography of Imprints and Publications, 1881–1890*. Reno: University of Nevada Press, 1981.

Asay, Jeff. *Western Pacific Timetables and Operations, A History and Compendium*. Crete, Neb.: J-B Publishing Company, 1983.

Ashbaugh, Don. *Nevada's Turbulent Yesterday: A Study in Ghost Towns*. Los Angeles: Westernlore Press, 1963.

Bailey, Edgar H., and Phoenix, David A. *Quicksilver Deposits in Nevada*. University of Nevada Bulletin vol. 38, no. 5. Reno: Nevada State Bureau of Mines and Mackay School of Mines, 1944.

Baker, E. Daniel. "Geology and Ore Deposits of the Bootstrap Subdistrict, Elko County, Nevada." *Geological Society of Nevada* 2, (1991): 619–623.

Bancroft, Hubert. *History of Nevada, Colorado and Wyoming, 1540–1888*. San Francisco: The History Company, 1890. Reprint, New York: Arno Press, 1967.

Bartlett, R. A. *Great Surveyors of the American West*. Norman: University of Oklahoma Press, 1962.

Basso, Dave. *Nevada Lost Mines and Hidden Treasures*. Sparks Nev.: Dave's Printing and Publishing, 1974.

Basso, Dave. *Nevada Historical Marker Guidebook: A Guide to 234 Historical Markers.* Sparks, Nev.: Falcon Hill Press, 1979.

Beal, Laurence H. *Investigation of Titanium Occurences in Nevada.* Nevada Bureau of Mines and Geology Report no. 3. Reno: Mackay School of Mines, University of Nevada, 1963.

Beebe, Lucius M. *The Central Pacific and the Southern Pacific Railroads.* Berkeley: Howell and North Books, 1963.

Beebe, Lucius, and Clegg, Charles. *U.S. West: The Saga of Wells Fargo.* New York: E. P. Dutton and Company, 1949.

Bell, Charles. *Nevada: Official Centennial Magazine.* Las Vegas: Charles Bell Publishing, 1964.

Bowman, Nora. *Only the Mountains Remain.* Caldwell, Idaho: Caxton Printers, 1958.

Boyd, Mrs. Orsemus. *Cavalry Life in Tent and Field.* Lincoln: University of Nebraska Press, 1982.

Buckley, E. R. "Geology of the Jarbidge Mining District." *Mining and Engineering World* 35 (1911): 1209–12.

Bushnell, Kent O. *Geology of the Rowland Quadrangle, Elko County, Nevada.* Nevada Bureau of Mines and Geology Bulletin no. 67. Reno: Mackay School of Mines, University of Nevada, 1967.

Carillo, F. V., and Price, J. G. *The Mineral Industry of Nevada, U.S. Bureau of Mines Minerals Yearbook: 1988.* Washington, D.C.: Government Printing Office, n.d.

Carlson, Helen S. *Nevada Place Names: A Geographical Dictionary.* Reno: University of Nevada Press, 1974.

Clawson, Marion. *From Sagebrush to Sage: The Making of a Natural Resource Economist.* Washington D.C.: Ana Publications, 1987.

Cline, Gloria Griffin. *Exploring the Great Basin.* Norman: University of Oklahoma Press, 1963.

Cloud, Barbara. *The Business of Newspapers on the Western Frontier.* Reno: University of Nevada Press, 1992.

Coash, John. R. *Geology of the Mount Velma Quadrangle, Elko County, Nevada.* Reno: Mackay School of Mines, University of Nevada, Nevada Bureau of Mines, 1972.

Coats, Robert R. *Geology of the Jarbidge Quadrangle.* USGS Bulletin 1141-M. Washington, D.C.: Government Printing Office, 1964.

———. *Geology of the Mountain City District.* Brigham Young University Geological Studies 2. Provo, Utah: Brigham Young University, 1964.

———. *Geology of Elko County, Nevada.* Nevada Bureau of Mines and Geology Bulletin no. 101. Reno: Nevada Bureau of Mines and Geology, University of Nevada, 1987.

Couch, Bertrand T., and Jay A. Carpenter. *Nevada's Metal and Mineral Production, 1859–1940.* University of Nevada Bulletin vol. 37, no. 4. Reno: University of Nevada, 1943.

Darrah, E. W. *Reviewing Nevada's Legacy.* Sepulvada, Calif.: The Sagebrush Press, 1964.

Davis, Sam P. *The History of Nevada.* Los Angeles: Elms Publishing Company, 1913.

Decker, Robert W. *Geology of the Bull Run Quadrangle, Elko County, Nevada.* Nevada

Bureau of Mines and Geology Bulletin no. 60. Reno: Mackay School of Mines, University of Nevada, 1962.

Denevi, Don. *Tragic Train, The City of San Francisco*. Seattle: Superior Publishing Company, 1977.

————. *The Western Pacific*. Seattle: Superior Publishing, 1978.

Douglas, I. H. "Geology and Gold and Silver Mineralization of the Tecoma District, Elko County, Nevada, and Box Elder County, Utah." Master's thesis, Stanford University, 1984.

Douglass, William A. *Basque Sheepherders of the American West*. Reno: University of Nevada Press, 1985.

Dunn, Hal, and Duane Feisel. *Nevada Trade Token Place Names*. Carson City, Nev.: Hal Dunn, 1973.

Egan, Howard. *Pioneering the West, 1846–1878*. Richmond, Utah: Howard R. Egan Estate, UCI Main Library, 1917.

Elko Chamber of Commerce. "Elko County the Agricultural Center of Nevada." Elko, Nev.: Free Press and Independent, n.d.

"Elko County and its Mineral Wealth." Elko, Nev.: Free Press Printing, n.d.

Elliott, Russell, and Helen Poulton. *Writings on Nevada: A Selected Bibliography*. Reno: University of Nevada Press, 1963.

Emmons, W. H. *A Reconnaissance of Some Mining Camps in Elko, Lander and Eureka Counties, Nevada*. USGS Bulletin no. 408. Washington, D.C.: Government Printing Office, 1910.

Farquhar, J. S. *Frémont in the Sierra Nevada*. Sierra Club Bulletin no. 15. N.p., n.d.

Federal Writers Project. *Nevada Towns*. Reno: n.p., 1940.

Ferguson, Henry. *The Mining Districts of Nevada*. Nevada Bureau of Mines Bulletin no. 40. Reno: University of Nevada, 1944.

Fletcher, N. F. *Early Nevada*. Reno: A. Carlisle and Company, 1929.

Florin, Lambert. *Ghost Towns of the West*. New York: Promontory Press, 1971.

Folkes, John. *Nevada Newspapers*. Reno: University of Nevada Press, 1964.

Fox, Theron. *Nevada Treasure Hunters Ghost Town Guide*. San Jose: Harlan-Young Press, 1961.

Frickstad, Walter, and Edward Thrall. *A Century of Nevada Post Offices*. Oakland, Calif.: Pacific Philatelic Research Society, 1958.

Friends of the Nevada Northern Railway. *Nevada Northern Railway and the Copper Camps of White Pine County, Nevada*. East Ely, Nev.: Friends of the Nevada Northern Railway, n.d.

Fulton, John A., and Alfred M. Smith. *Nonmetal Minerals in Nevada*. University of Nevada Bulletin no. 26. Reno: University of Nevada, 1932.

Galloway, John. *The First Transcontinental Railroad, Central Pacific, Union Pacific*. New York: Simmons-Boardman, 1950.

Gammett, James, and Stanley Paher. *Nevada Post Offices*. Las Vegas: Nevada Publications, 1989.

Gianella, Vincent P. Barite Deposits of Northern Nevada." *AIME Mining Technology* 4 (1940).

Gillerman, V. S. *Geology of the Railroad (Bullion) District*. N.p.: Association of Exploration Geochemists, 1984.

Goodwin, Victor. *The Humboldt, Nevada's Desert River and Thoroughfare of the American West.* Washington, D.C.: U.S. Department of Agriculture, Nevada River Basin Survey, 1966.

Granger, Arthur E., Mendell M. Bell, George C. Simmons, and Florence Lee. *Geology and Mineral Resources of Elko County.* Nevada Bureau of Mines and Geology Bulletin no. 54. Reno: University of Nevada, 1957.

Greenwell, Glen H. *A Last Look.* Beckwourth, Calif.: Louise Deserio, 1979.

Grover, David. *Diamondfield Jack, A Study in Frontier Justice.* Reno: University of Nevada Press, 1968.

Hanks, Edward. *A Long Dust in the Desert.* Sparks, Nev.: Western Printing and Publishing, 1967.

Hanley, Mike. *Owyhee Trails, the West's Forgotten Corner.* Caldwell, Idaho: The Caxton Printers, 1975.

Harris, Robert P. *Nevada Postal History.* Santa Cruz: Bonanza Press, 1973.

Haws, Adelaide. *Valley of Tall Grass.* Bruneau, Idaho: Caxton Publishers, 1950.

Higgins, James, Eric Moody, and Lee Mortensen. *A Preliminary Checklist of the Manuscript Collections at the Nevada Historical Society.* Reno: Nevada Historical Society, 1974.

Higgs, Gerald B. *Lost Legends of the Silver State.* Salt Lake City: Western Epics, Inc., 1976.

Hill, James M. *Notes on Some Mining Districts in Eastern Nevada.* USGS Bulletin no. 648. Washington, D.C.: Government Printing Office, 1916.

Hitt, Douglas. *The Original Ghost Town Directory.* Carson City, Nev.: Ghost Towns, Limited, 1970.

Holbrook, Marjorie. *History of Metropolis, Nevada.* Marjorie Holbrook, 1986.

Howard, Robert West. *The Great Iron Trail.* New York: S. P. Putnam and Sons, 1883. Reprint, New York: Putnam, 1962.

Johnson, Maureen G. *Placer Gold Deposits of Nevada.* USGS Bulletin no. 1356. Washington, D.C.: Goverment Printing Office, 1973.

Johnson, Robert Neil. *California-Nevada Ghost Town Atlas.* Susanville, Calif.: Cy Johnson and Son, 1970.

Jones, Richard B. *Directory of Nevada Mining Operations Active During 1985.* Reno: Nevada Bureau of Mines and Geology, 1985.

Ketner, Keith B., and J. Fred Smith Jr. *Geology of the Railroad Mining District, Elko County, Nevada.* USGS Bulletin no. 1162-B. Washington, D.C.: Goverment Printing Office, 1963.

Klepper, M.R. *Star Tungsten Mine.* USGS Open File Report. Washington D.C.: Government Printing Office, 1943.

Kneiss, Gilbert. *Bonanza Railroads.* Stanford: Stanford University Press, 1941.

Koontz, John. *Political History of Nevada.* Carson City, Nev.: State Printing Office, 1960.

Kraus, George. *High Road to Promontory: Building the Central Pacific Across the High Sierra.* Palo Alto: American West Publishing Company, 1969.

LaPointe, Daphne D., Joseph V. Tingley, and Richard B. Jones. *Mineral Resources of Elko County, Nevada.* Nevada Bureau of Mines Bulletin no. 106. Reno: Mackay School of Mines, University of Nevada, 1991.

Larrison, Earl. *Owyhee: The Life of a Northern Desert*. Caldwell, Idaho: Caxton Printers, 1957.

Lavender, David. *The Great West*. Palo Alto: American Heritage Publishing Company, 1965.

Laxalt, Robert. *Sweet Promised Land*. New York: Harper and Row, 1957. Reprint, Reno, University of Nevada Press, 1986.

———. *In a Hundred Graves, A Basque Portrait*. Reno: University of Nevada Press, 1972.

———. *Nevada: A History*. W. W. Norton and Company, 1977. Reprint, Reno: University of Nevada Press, 1991.

Leigh, R. W. *Nevada Place Names: Their Origin and Significance*. Salt Lake City: Deseret News Press, 1964.

Lincoln, Francis Church. *Mining Districts and Mineral Resources of Nevada*. Nevada Newsletter Publishing Company, 1923.

Lingenfelter, Richard E., and Karen Rix Gash. *The Newspapers of Nevada: A History and Bibliography, 1854–1979*. Reno: University of Nevada Press, 1984.

Loofbourow, Leon. *Steeples Among the Sage*. Oakland: University of the Pacific, 1964.

Lovering, T. S., and W. M. Stoll. *Preliminary Report on the Rip Van Winkle Mine*. USGS Open File Report. Washington, D.C.: Government Printing Office, 1943.

Mack, Effie Mona. *Nevada*. Glendale, Calif.: Arthur H. Clark, 1936.

Marshall, Howard, and Richard Ahlborn. *Buckaroos in Paradise: Cowboy Life in Northern Nevada*. Library of Congress, 1980.

Mathias, Donald. *I'd Rather Be in Jarbidge*. Glendora, Calif.: Dojeri Publications, 1986.

McClellan, E. C. *Elko County: Location and Site and a Full Description of its Agricultural Mineral and Quarry Resources and Climate and Rainfall*. Elko, Nev.: Independent Job Print, 1891.

McDonald, Douglas. *Nevada Lost Mines and Buried Treasures*. Las Vegas: Nevada Publications, 1981.

McElrath, Jean. *Aged in Sage*. Reno: University of Nevada Press, 1964.

———. *Tumbleweeds, 1940–1967*. Reno: University of Nevada Press, 1971.

McKinney, Whitney. *A History of the Shoshone-Paiutes of the Duck Valley Indian Reservation*. Salt Lake City: The Institute of the American West and Howe Brothers, 1983.

Metal and Non-metal Occurrences in Nevada. University of Nevada Bulletin vol. 26, no. 6. Reno: University of Nevada, 1932.

Miller, Lester W. *A Sagebrush Saga*. Springfield, Utah: Art City Publishing Company, 1956.

Mitchell, James R. *Gem Trails of Nevada*. Baldwin Park, Calif.: Gem Guides Book Company, 1991.

Moody, Eric. *An Index to the Publications of the Nevada Historical Society, 1907–1971*. Reno: Nevada Historical Society, 1977.

Morrisey, Frank R. *Turquoise Deposits of Nevada*. Nevada Bureau of Mines and Geology Report no. 17. Reno: Mackay School of Mines, University of Nevada, 1968.

Murbarger, Nell. *Ghost of the Glory Trail*. Palm Desert, Calif.: Desert Magazine Press, 1956.

———. *Sovereigns of the Sage*. Palm Desert, Calif.: Desert Magazine Press, 1958.

Myrick, David. *Railroads of Nevada and Eastern California*. 2 vols. Howell-North, 1963. Reprint, Reno: University of Nevada Press, 1992.

Nevada: The Silver State. Carson City, Nev.: Western States Historical Publishing, 1970.

Nevada Department of Economic Development. *Nevada Community Profiles*. Carson City, Nev.: State Printing Office, 1964.

Nevada Writers Project. *Nevada*. Carson City, Nev.: Binfords and Mort, 1940.

Nielson, Norm. *Tales of Nevada*. Vols. 1 and 2. Reno: Tales of Nevada, 1989.

Nolan, Thomas B. *The Tuscarora Mining District, Elko County, Nevada*. Nevada Bureau of Mines and Geology Bulletin no. 25. Reno: Nevada State Bureau of Mines and Mackay School of Mines, University of Nevada, 1936.

Norman, Loyal Vernon. *A Slice of Nevada School Reorganization*. Philadelphia: Dorrance Publishing, 1964.

O'Bryan, Frank. *Overland Chronicle: Emigrant Diaries in Western Nevada Libraries*. Reno: Nevada Historical Society, n.d.

Owens, Preston J. *Wes Helth: A Man For All Reasons*. Provo, Utah: Brigham Young University Family History Services, 1990.

Paher, Stanley. *Nevada Ghost Towns and Mining Camps*. Las Vegas: Nevada Publications, 1970.

———. *Nevada Official Bicentennial Book*. Las Vegas: Nevada Publications, 1976.

———. *Nevada, An Annotated Bibliography*. Las Vegas: Nevada Publications, 1980.

———. ed. *Nevada Towns and Tales, Volume 1, North*. Las Vegas: Nevada Publications, 1981.

Papke, Kieth G. *Fluorspar in Nevada*. Nevada Bureau of Mines and Geology Bulletin no. 93. Reno: Mackay School of Mines, University of Nevada, 1979.

———. *Barite in Nevada*. Nevada Bureau of Mines and Geology Bulletin no. 98: Reno: Mackay School of Mines, University of Nevada, 1984.

Pardee, J.T., and E.L. Jones. *Deposits of Manganese Ore in Nevada*. USGS Bulletin 710-F. Washington, D.C.: Government Printing office, 1920.

Paris, Beltran, with William A. Douglass. *Beltran, Basque Sheepman of the American West*. Reno: University of Nevada Press, 1979.

Park, John. *Jarbidge*. USBM Information Circular no. 6543. Washington, D.C.: Government Printing Office, 1931.

Parker, Carlyle P., and Janet Parker. *Nevada Biographical and Genealogical Sketch Index*. Turlock, Calif.: Marietta Publishing, 1986.

Patterson, Edna B. *Who Named It?* Elko, Nev.: Elko Independent, 1965.

———. *Sagebrush Doctors*. Springville, Utah: Art City Publishing, 1972.

———. *This Land Was Ours*. Springville, Utah: Art City Publishing, 1973.

———. *Indian Paint Brush*. Springville, Utah: Art City Publishing, 1982.

Patterson, Edna B. and Louise A. Beebe. *Halleck Country, Nevada: The Story of the Land and Its People*. Reno: College of Agriculture, University of Nevada, 1982.

Patterson, Edna, Louise Ulph, and Victor Goodwin. *Nevada's Northeast Frontier*. Reprint, Reno: University of Nevada Press, 1991.

Peacock, E. *Report on the Cornucopia Group of Mines and the Leopard Mine*. Nevada Bureau of Mines and Geology, Mining District File no. 49. Reno: Nevada Bureau of Mines and Geology, 1922.

Penfield, Thomas. *A Guide to Treasure in Nevada.* Conroe, Tex: True Treasurer, 1974.

Ramsey, Robert H. *Men and Mines of Newmont: A Fifty-Year History.* New York: Octagon Books, 1973.

Reed, Flo. *Bygone Days of Nevada Schools.* Flo Reed, 1991.

Rott, E.H., Jr. *Ore Deposits of the Gold Circle Mining District, Elko County, Nevada.* Nevada Bureau of Mines Bulletin no. 12. Reno: University of Nevada, 1931.

Sawyer, Byrd. *The Gold and Silver Rushes of Nevada.* San Jose: Harlan-Young Press, 1971.

Schilling, John. *The Nevada Mineral Industry, 1980.* Nevada Bureau of Mines and Geology Special Publication MI-1980. Reno: University of Nevada.

————. *The Nevada Mineral Industry, 1981.* Nevada Bureau of Mines and Geology Special Publication MI-1981. Reno, University of Nevada, 1981.

————. *The Nevada Mineral Industry, 1985.* Nevada Bureau of Mines and Geology Special Publication MI-1985. Reno: University of Nevada, 1985.

————. *The Nevada Mineral Industry, 1990.* Nevada Bureau of Mines and Geology Special Publication MI-1990. Reno: University of Nevada, 1990.

Scott, Kenneth. "Calvacade of Time." *Elko Independent,* 1982.

Scrugham, James G. *Nevada, a Narrative of the Conquest of a Frontier Land: Comprising the Story of Her People from the Dawn of History to the Present Time.* Chicago: The American Historical Society, 1935.

Shawe, Fred R., Robert G. Reeves, and Victor E. Kral. *Iron Deposits of Nevada, Part C, Iron Ore Deposits of Northern Nevada.* Nevada Bureau of Mines and Geology Bulletin no. 53. Reno: University of Nevada, 1962.

The Silver Mines of Nevada. New York: William C. Bryant and Company, 1864.

Smith, Alfred M. "Mountain City District's Mines of Copper, Gold and Silver." *Mining Review* 34 (1932): 6–7.

Smith, Alfred M., and William O. Vanderburg. *Placer Mining in Nevada.* University of Nevada Bulletin vol. 26, no. 8. Reno: University of Nevada, 1932.

Smith, Raymond. *Saloons of Old and New Nevada.* Minden, Nev.: Silver State Printing, 1992.

Smith, Roscoe. *Mineral Resources of Elko County.* USGS Open File Report no. 76–56. Washington, D.C.: Government Printing Office, 1976.

Stager, Harold K., and Joseph V. Tingley. *Tungsten Deposits in Nevada.* Nevada Bureau of Mines and Geology Bulletin no. 105. Reno: Nevada Bureau of Mines and Geology, University of Nevada, 1988.

Stapley, Linda. "Together in My Name: A Centennial Tribute to the Little Church of the Crossroads, Lamoille, Nevada." *Elko Independent,* 1990

State Mineralogist of Nevada. *Annual Reports, 1864–1928.* Carson City, Nev.: State of Nevada.

Steininger, Roger. *Geology of the Kingsley District.* Brigham Young Geological Studies no. 13. Provo, Utah: Brigham Young University, 1966.

Stevens, Horace. *The Copper Handbook: A Manual of the Copper Industry of the United States.* Vols. 2–10. Washington, D.C.: Government Printing Office, 1908.

Sylvester, James. *Ferber Mining District.* Salt Lake City: University of Utah, 1950.

Truett, Velma. *On The Hoof in Nevada.* Los Angeles: Gehrett-Truett-Hall, 1950.

Tuscarora Stamp Mills. Nevada Bureau of Mines Bulletin no. 54. Reno: University of Nevada, 1957.

Vanderbilt, Paul. *Guide to the Special Collections of Prints and Photographs in the Library of Congress.* Washington, D.C.: Library of Congress, 1955.

Vanderburg, William O. *Placer Mining in Nevada.* University of Nevada Bulletin 30, no. 4. Reno: University of Nevada, 1936.

Van Meter, David L. *G. S. Garcia, Elko, Nevada: A History of the World Famous Saddlemaker.* Reno: Avail Publishing, 1984.

Various Authors. *Ruby Valley Memories.* N.p.: n.d.

Weed, Walter. *The Copper Handbook: A Manual of the Copper Mining Industry of the World.* Vol. 2. Washington, D.C.: Government Printing Office, 1912.

————. *The Mines Handbook and Copper Handbook, 1916–1926.* Washington, D.C.: Government Printing Office, 1927.

Wilson, Clark. "The Geology of Black Forest Mine: Spruce Mountain, Nevada." 1938, Northeastern Nevada Museum manuscript collection, Elko, Nevada.

Wilson, Helen. *Gold Fever.* La Mesa, Calif.: Helen Wilson, 1974.

Winchell, Bessie. *Now and Then.* Prineville, Oreg.: Bonanza Publishing, 1986.

Works Progress Administration. *Inventory of the County Archives of Nevada, No. 4 Elko County.* Works Progress Administration, 1938.

Wren, Thomas. *The State of Nevada: Its Resources and People.* New York: Lewis Publishing Company, 1904.

Manuscripts

Allen, Vicki. "Tuscarora."

Blazek, Penny. "The Mysteries of Lone Mountain, Nevada."

Brennen, Thomas. "Life at Fort Halleck."

Bruner, Victor. "Reminiscences of Early Elko and Elko County, Nevada."

Burns, Linda. "Lamoille."

Butler, Julie. "Tuscarora Chinese."

Chapin, Kevin. "Tuscarora."

Christean, Teryl Ann. "John Jesse Hylton."

Copeland, Teresa. "Jarbidge."

Davidson, P. W. "History of Early Jarbidge."

Davis, James. "The Gallagher Fish Hatchery."

Evans, Patty. "Transportation in Elko County."

Garteiz, Dennis. "Tuscarora Boom Days."

Hackett, Eddie. "The History of the Rio Tinto Mine."

Harrison, Mavis. "Elko County History."

Hart, Rod. "Staging Enterprises of Nevada."

Heguy, Lori. "The Six-toed Murderer."

Hickson, Howard. "Place Names of Northern Nevada."

Hurley, Lulu Belle. "The History of the Development of Elko County."

Larson, Linda. "Old Allegheny."

Martin, Hugh. "Northern Elko County."

Mattice, Kathleen. "Recollections of Ruby City."

Moore, Susan. "John Jesse Hylton."

Moschetti, Ann. "Midas."

Palacio, Robert. "Ruby City."

Paoletti, Pat. "Metropolis."

Patterson, Edna. "Fort Halleck."

Roseberry, Roy. "Tuscarora Mining District."

Scott, Stephanie. "Tuscarora."

Sewell, Carol. "Tuscarora Time."

Smith, Carol. "A tour of the Utah Construction Company."

Uhlig, Jennifer. "Metropolis."

Westlund, Linda. "History of the Lamoille Church."

Newspapers and Periodicals

Business Talks (Tuscarora)

Carlin Express

Carlin Courier

Commonwealth (Carlin)

Commonwealth (Deeth)

Contact Miner

Contact News

Daily Argonaut (Elko)

Elko Chronicle

Elko Enterprise

Elko Independent

Elko Post

Engineering and Mining Journal

Free Press (Elko)

Gold Circle Miner (Midas)

Gold Circle News (Midas)

Gold Circle Porcupine (Midas)

Gold Creek News

High Desert Advocate (Wendover)

Metropolis Chronicle

Mining and Scientific Press

Mining News (Tuscarora)

Mining Review (Tuscarora)

Mountain City Mail

Mountain City Times

Nevada Democrat (Carlin)

Nevada Historical Society Quarterly

Nevada Silver Tidings (Elko) (Deeth)

Nevada State Herald (Wells)

Northeastern Nevada Historical Society Quarterly

Rio Tinto News

Ruby Valley News

Telegram (Elko)
Tobar Times
Tuscarora Mining News
Tuscarora Times
Tuscarora Times-Review
Wells Progress
Western Home Builder (Carlin)

Personal Communications

W. E. Clawson Jr.
Bob Erickson
Morris Gallagher
Nelda Glaser
Michelle Gonzalez
Elizabeth Pruitt

Index